ANALYZING NONRENEWABLE RESOURCE SUPPLY

ANALYZING NONRENEWABLE RESOURCE SUPPLY

Douglas R. Bohi
Michael A. Toman

Resources for the Future / Washington, D.C.

ACKNOWLEDGMENT OF SUPPORT AND LEGAL NOTICE

Analyzing Nonrenewable Resource Supply was financed by the Electric Power Research Institute (EPRI) of Palo Alto, California under Research Project 1220-1. The findings and conclusions, however, are those of the authors and not of EPRI or any of its member organizations.

In keeping with its policies, EPRI also makes the following legal notice:

"This report was prepared by Resources for the Future as an account of work sponsored by the Electric Power Research Institute, Inc. ("EPRI"). Neither EPRI, members of EPRI, nor Resources for the Future, nor any person acting on behalf of either:

"a. Makes any warranty or representation, express or implied, with respect to the accuracy, completeness, or usefulness of the information contained in this report, or that the use of any information, apparatus, method, or process disclosed in this report may not infringe privately owned rights; or

"b. Assumes any liabilities with respect to the use of, or for damages resulting from the use of, any information, apparatus, method, or process disclosed in this report."

Manufactured in the United States of America

Published by Resources for the Future, Inc., 1755 Massachusetts Avenue, N.W.
Washington, D.C. 20036
Resources for the Future books are distributed worldwide by The Johns
Hopkins University Press

Library of Congress Cataloging in Publication Data

Bohi, Douglas R.
Analyzing nonrenewable resource supply.

Bibliography: p.
Includes index.
1. Energy industries. 2. Power resources—Economic aspects. 3. Supply and demand. 4. Energy policy.
I. Toman, Michael A. II. Resources for the Future.
III. Title.
HD9502.A2B66 1984 333.79'11 83-43264
ISBN 0-915707-05-5
ISBN 0-915707-06-3 (pbk.)

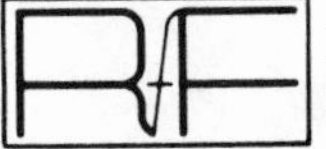

Resources for the Future is a nonprofit organization for research and education in the development, conservation, and use of natural resources, including the quality of the environment. It was established in 1952 with the cooperation of the Ford Foundation. Grants for research are accepted from government and private sources only on the condition that RFF shall be solely responsible for the conduct of the research and free to make its results available to the public. Most of the work of Resources for the Future is carried out by its resident staff; part is supported by grants to universities and other nonprofit organizations. Unless otherwise stated, interpretations and conclusions in RFF publications are those of the authors; the organization takes responsibility for the selection of significant subjects for study, the competence of the researchers, and their freedom of inquiry.

This book was prepared in RFF's Center for Energy Policy Research, Joel Darmstadter, director. It was edited by F. R. Ruskin and designed by Elsa B. Williams. The index was prepared by Florence Robinson.

Contents

Foreword

Confusion about the determinants of nonrenewable resource supply is one of the vexing problems at the intersection of economics and public policy. Specifically, questions of oil and natural gas supply response have been at the core of arguments over price controls, taxes, and other policy measures. Supply responses also figure prominently in concern about future availability and cost of petroleum and minerals.

What is needed to clear up the confusion is an approach that brings together a theory of resource supply, data to make that theory operational, and a methodology to bring the two together—all set within an institutional framework sufficiently rich to illuminate behavioral consequences. Without this amalgam, economists will not be able to answer questions of the sort suggested above with sufficient precision and certainty to provide specific and persuasive policy guidance. In a broader sense, their conclusions on resource supply trends over the longer term will remain conjectural.

In market economies such as the United States, approaches grounded in abstract social decision rules have limited relevance. An adequate theory must start with the behavior of individual actors and the interaction of incentives, expectations, and decisions over time. Moreover, the process of bringing resources to the market integrates across production stages and cannot be successfully modeled otherwise. Another issue that deserves special attention is the aggregation of individual behavior at the industry level.

Bohi and Toman have approached the modeling of supply behavior from this dynamic and market-oriented perspective. In particular, the reader will not find here the omniscient social decision maker optimizing across time. Moreover, their analysis is not concerned with theory alone but with the practical problems of applying the theory with imperfect data. As the authors make clear, understanding resource supply behavior requires a unified treatment of concepts, facts, and methodology.

Starting from this vision of what a useful model of supply behavior would entail, the authors examine the literature to see what it offers. They find much of what is required missing, but the cupboard is not bare. And along the way, they provide many of the pieces to bridge the gaps. Just as important, their careful reconnaissance highlights the missing elements and directs attention to the research that is required if models of supply behavior are to be useful.

This careful work provides a caution, a guide, and a framework for further research. The caution is against uncritical acceptance of what sometimes appear to be answers to supply responses of nonrenewable resources. The guide is to what is known. Those who build models of supply can determine how, even with the current state of knowledge, productive work can be done, and those who rely upon such models can discern the limits of insights that can fruitfully be obtained. The research framework describes the efforts needed to broaden the range of those insights and to make them more precise. Consequently, this book will be useful to students, to practitioners, and to those who need to know more about the interaction of economic forces and the resource base.

Washington, D.C., June 1983

Milton Russell
Director, Center for
Energy Policy Research

Preface

This study is concerned with the general problem of using economic models of nonrenewable resource supply to understand actual supply behavior. Our motivation for preparing the volume arises from several sources. At the broadest level, it stems from the fragmentation of the literature on nonrenewable resources into separate treatments of theory, methodology, institutions, and data, with few attempts to draw the strands together. Theoretical analyses are seldom concerned with the practical problems of deriving empirical models amenable to hypothesis testing and prediction; conversely, empirical studies tend to be *ad hoc* and only distantly related to theoretical fundamentals. This volume is an attempt to draw these strands together and to bridge the gap between theory and application, with an emphasis on applications to petroleum supply behavior.

A second and related source of motivation is the generally poor performance of empirical models in explaining observed supply behavior, particularly the behavior of oil and natural gas supply in the United States. Overall, the statistical findings have been only moderately successful in explaining historical behavior and notably unsuccessful in predicting petroleum supply activities in response to changes in economic incentives and regulatory constraints. This poor performance has contributed to the view that economic analysis has little to offer in understanding actual behavior. In our view, the poor record of empirical studies may be explained in part by the gap between theory and application referred to above.

The third motive arises from the observation that economists and other analysts place undue emphasis on the so-called *r* percent rule, and on the narrow context in which this rule is valid, when investigating both conceptual and empirical issues. These analyses are dangerously close to empty economic boxes, yet unnecessarily so because the theory has been developed far beyond these narrow strictures.

Finally, there have been several major advances in the theory of nonrenewable resources in recent years, and developments in empirical methods for analyzing dynamic economic behavior generally and nonrenewable resources in particular, that deserve to be drawn together. Interest in the determinants of energy supply has stimulated numerous conceptual and empirical studies of resource supply behavior. These advances have extended the frontiers of research well beyond the limits reached only a decade ago. Nevertheless, critical deficiencies remain, and these deficiencies deserve highlighting as well.

Much of the study is a reinterpretation and evaluation of existing research, though the dual focus on conceptual and practical questions distinguishes the work from other studies that treat these issues in isolation. We present no new empirical estimates of supply (a formidable undertaking in its own right), nor, for the most part, do we present fundamentally new theoretical models and findings. Exceptions include the analysis of cost specifications in the appendix to chapter 2; the critique of game theory models applied to common property externalities in chapter 3; and the discussion of uncertainty, rational expectations, and "certainty equivalence" in chapter 4 and its appendix. In a work of this scope, it is also inevitable that some topics receive less emphasis than others. One important subject that we do not treat in any detail is the implication of various contracting or bidding procedures for mineral rights allocation and for long-term supply relationships in industries such as coal and natural gas.

Throughout the study we have attempted to focus on intuitively understandable concepts while avoiding rigorous but lengthy technical arguments, which would divert attention from the main subject. References to more formal treatments are supplied in both the text and the footnotes. Our intent has been to produce a volume with an appeal to a wide variety of readers. We hope that it provides a valuable reference and commentary to developers and users of empirical resource supply models (particularly energy models) in academic, government, and business circles. Students of resource economics, dynamic economics, and applied econometrics also may find it a useful tool. Finally, we hope that the study is of general interest to economists and other analysts concerned with empirical applications of theoretical constructs.

Our work on this volume was funded by a grant to Resources for the Future from the Electric Power Research Institute (EPRI). We sincerely appreciate the assistance and counsel of the late Albert N. Halter, project manager at EPRI through most of the work period. We are also indebted to Michael Bowes, Ronald Cummings, Dennis Epple, Richard Gilbert, and other, anonymous reviewers for reading different drafts of the manuscript and providing helpful comments; and to W. David Montgomery, Robert Pindyck, and Milton Russell for helpful discussions on various aspects of the subject. Final responsibility for the work rests, of course, with us alone. A large intellectual debt is owed to the many persons who have contributed to the literature on nonrenewable resources. We hope we have done justice to the contributions.

We have received valuable assistance from Caroline Bouhdili in proofreading the manuscript, providing references, and drawing figures, and from Sandra Glatt and Steve Polasky for their help with an earlier version. Anne Farr and Rose Mongoba typed endless earlier drafts, while Nancy Waldo prepared the final version. To all of these persons we are deeply indebted.

Washington, D.C., May 1983

Douglas R. Bohi
Michael A. Toman

1
Introduction

This study provides a unified treatment of the principal theoretical and empirical issues confronted in the analysis of nonrenewable resource supply behavior in a market-oriented economy. By a unified treatment, we refer to the three dimensions of analysis—theory, data, and estimation method—that must be combined to successfully explain and predict supply behavior. The theory provides hypotheses that may be tested against actual observations using empirical techniques. These hypotheses should be based upon plausible descriptions of the supply process, should offer unambiguous implications about behavior, and should require information that matches available data. The empirical methodology in turn should suitably match the requirements of the theory and data.

Advances in theoretical and empirical methods during the last decade have improved our ability to understand the supply process. Nevertheless, serious deficiencies remain. All too often, theoretical research is conducted without sufficient regard for its empirical content. Similarly, empirical analyses are frequently based on *ad hoc* models of supply behavior that are only distantly related to the theory. One purpose of this volume is to assess current research on nonrenewable resources by describing its limits and the deficiencies that remain.

The major part of the book (chapters 2 through 5) covers theoretical issues concerning the supply of nonrenewable resources in a fairly general setting. Attention is focused on the plausibility and completeness of conceptual models of supply, on the behavioral implications of these models, and on the problems of using these models to describe actual behavior. The matter of consistency among theory, data, and empirical methods is illustrated by reference to the U.S. petroleum industry in chapter 6. Data on costs, prices, and supply activities for this industry are compared with their counterparts in the theory, and statistical methods of estimating oil and natural gas supply are compared with conceptual descriptions of the supply process. Directions for further research are suggested in chapter 7.

The scope of the material included here is limited to that of supply decisions made by individual profit-motivated firms in a market economy.[1] Specifically excluded are models of socially optimal use of a fixed resource stock in a central planning context.[2] Models in this category are not particularly useful for understanding the behavior of private firms because they abstract from the many factors that impinge on individual decisions and from the interactions between supply and demand in a market economy.

All of the conceptual models examined in this study predict supply behavior of firms as the solution to an intertemporal maximization problem. Firms are presumed to choose paths of resource extraction and to make investment decisions that maximize the present value of net revenue (possibly adjusted for risk) given current prices and costs and expectations of future influences. The specifics of each model vary with the category of behavior being considered (for example, extraction versus exploration decisions), but the emphasis is on the dynamic response of supply behavior to economic influences. As indicated in the next section, these responses are an integral part of nonrenewable resource supply behavior.

The objects of interest in the conceptual analysis are the decision rules implied by the maximization problem, rather than the details of the production processes themselves. Thus, the models abstract from physical and engineering relationships by employing the standard procedure in economic theory of representing production processes by cost and production functions. Similarly, the models abstract from considerations pertaining to the geological distribution and geophysical properties of resource deposits and from the issues that arise in estimating the size of

1. This may be expanded to include supply behavior in economies where mineral rights are publicly owned and are leased to profit-motivated firms that make exploration and production decisions on the basis of economic considerations.

2. See Dasgupta and Heal (1979) for a discussion of these models.

the "resource base" and the physical "scarcity" of nonrenewable resources. Finally, the peculiar institutional characteristics of various resource industries, particularly the oil and natural gas industries, are discussed only insofar as they relate to broader questions concerning analysis of nonrenewable resource markets and regulatory constraints.[3]

As an introduction to the subject, the remainder of this chapter describes characteristics typical of the supply process for nonrenewable resources and the salient issues in understanding how the process works.

CHARACTERISTICS OF THE SUPPLY PROCESS

Understanding the behavior of nonrenewable resource supply is complicated by the fact that the process of supply consists of several overlapping sets of activities rather than a single transformation of inputs to outputs. Decisions at each stage of the process, and the interactions among these decisions, must be analyzed to understand the process as a whole. Moreover, decisions at each stage depend on intertemporal considerations: at any point in time, the profitability of decisions depends on current prices and costs, on past decisions as influenced by past prices and costs, and on anticipated future prices and costs. The fact that future prices and costs are uncertain introduces the issue of how firms respond to risk and how expectations are determined.

The Stages of Supply

Transforming nonrenewable resources in the ground to extracted commodities ready for sale involves three consecutive, overlapping stages: exploration, development, and extraction. Exploration involves determining the location of potential deposits and estimating their geophysical traits. This stage shades over into development, which involves additional delineation of deposits and preparation of sites for extracting resources. Development in turn shades over into the extraction stage of the process. Each of these stages consists of several distinct activities.[4]

3. For an in-depth discussion of these issues as they relate to the U.S. petroleum and coal industries, see McDonald (1971) and Zimmerman (1975), respectively. Nehring (1981) provides an example of the methodological procedures and concerns which arise in estimating a resource base, while the "scarcity" issue is examined from several perspectives in the papers collected by Smith (1979) and Smith and Krutilla (1982).

4. See McDonald (1971) and Zimmerman (1975) for details. In the United States, some firms specialize in one stage (for example, independent petroleum exploration firms), while others are fully integrated into all stages. This aspect of market structure has implications for the use of available data to analyze supply behavior, as indicated in chapter 5.

Other stages in the supply process include, for hard-rock minerals, milling, refining, and processing after extraction and, for natural gas, the maintenance of above-ground inventories and connections to a pipeline distribution system. In countries where resource-bearing lands are privately owned or where rights to exploit the mineral potential of public lands are allocated by a market mechanism, the acquisition of resource-bearing lands and mineral rights is another important stage in the supply process. With the exception of land acquisition, however, these activities are not subject to the inherent effects of depletion which distinguish nonrenewable resources. The focus of this study is on the exploration, development, and extraction stages of the process where these effects appear.

Development and exploration can be further divided into activities on the "intensive margin" and on the "extensive margin." Development on the intensive margin involves installing more capital in a given producing area to increase its yield, while development on the extensive margin refers to the preparation of new sites for extraction. Similarly, exploration on the intensive margin refers to the search in unexplored areas immediately adjacent to, or structurally similar to, known deposits; exploration on the extensive margin refers to the search in new and unfamiliar territories. In practice, the distinctions between these margins are somewhat arbitrary because the data often do not differentiate between exploration and development or between extensive and intensive activities. Nevertheless, the broad differences between exploration and development activities are important in empirical analyses of supply behavior.

The decisions made at each stage of the supply process—exploration, development, and extraction—are intricately interconnected. The output of exploration satisfies a "derived demand" for inputs to development, and the output of development satisfies a derived demand for inputs to extraction. New discoveries and new reserves also ameliorate (for a time) the effects of depletion. New discoveries offset rising development costs, and newly developed reserves offset rising extraction costs.[5] The connections imply that changes in economic or institutional factors influencing behavior at one stage will also influence behavior at other stages. For example, changes in the price of final output will alter exploration and development decisions as well as extraction behavior; changes in the cost

5. Depletion effects can also be ameliorated by technical progress, as illustrated by Fisher (1964, pp. 53–57). Following Kamien and Schwartz (1969), technical change can be thought of (at least in part) as the response to rising resource prices and costs, just as rising extraction cost stimulates new reserve development and rising development cost stimulates exploration.

or value of new reserves will alter extraction decisions as well as exploration and development activity.

Depletion Effects and Resource Scarcity

The distinguishing characteristic of nonrenewable resources, and therefore the heart of conceptual descriptions of supply behavior, is the effect of current depletion on the cost of future activities. Within a single petroleum reservoir, for example, current extraction reduces reservoir pressure and raises the cost of extraction in subsequent periods. To a first approximation, therefore, current extraction cost can be regarded as positively correlated with the cumulative volume of past extraction or, equivalently, as negatively correlated with the volume of remaining reserves (Uhler, 1979). Similarly, the cost of extraction from an underground coal mine increases with the cumulative volume of past extraction as the mine face becomes increasingly distant from the mine head and as the richest seams are extracted first (Zimmerman, 1977).

Depletion effects also arise in development and exploration. Development cost eventually rises as a resource deposit is worked more intensively because of scale diseconomies (such as well interference as spacings become narrower) and because of declining resource quality (that is, lower grade deposits). Development costs also rise on the extensive margin, as the best sites among an inventory of known deposits normally are developed first (Solow and Wan, 1976). These depletion effects can be thought of as being positively correlated with the cumulative volume of reserves developed, or as being negatively correlated with the stock of discovered reserves remaining to be developed. Similar relationships are encountered in exploration: on average, the cost of new discoveries rises over time (especially on the extensive margin) with the volume of cumulative discoveries as the most attractive prospects are explored first (Peterson, 1978; Nehring, 1981).[6]

Costs tend to rise over time because there are only finite stocks of nonrenewable resources present in the earth's crust. However, it is not this geological finiteness per se that is important in modeling supply behavior; rather, it is the rising cost trends *implied* by the finiteness of the natural endowment. Rising costs will restrict resource supply, cause the resource price to rise, and force consumers to conserve on the use of the resource. If the price rises high enough, consumption will fall to zero before the natural endowment is physically exhausted. The link between depletion effects on the supply side and conservation on the demand side

6. The caveat "on average" is particularly important at the exploration stage because of the inherent uncertainties in finding resources and because of certain spillover effects that can result from these uncertainties (Peterson, 1978).

is strengthened if an inexhaustible substitute becomes feasible at some price. Examples are solar, wind, and fusion power as substitutes for fossil fuels.

Despite the fact that resources are nonrenewable, the constraints on cumulative resource production imposed by physical exhaustion need not, and likely will not, be binding. Consequently, in most of this book it is assumed that resource supply is constrained by rising costs rather than by ultimate physical limits. The distinction, which is expanded upon in the next chapter, emphasizes "economic exhaustion" rather than physical exhaustion ("running out").

THE DYNAMICS OF SUPPLY

The effect of depletion on production costs has important implications for modeling supply behavior under the assumption that firms seek to maximize the present value of discounted net revenues. Firms will make extraction decisions according to the balance between the current gain in revenue and the added cost of increased extraction, where the added cost includes current operating cost and the present value of lower future profits because depletion raises future costs. Similarly, firms will make development and exploration decisions according to the balance between the increase in present value from larger reserves and the added cost of new reserves. Here the added cost includes current investment costs and the present value of lower future profits because of depletion-induced costs.

Expectations about future profits depend not only on depletion but also on future resource prices and on exogenous influences on costs (including changes in the technology of finding and producing the resource). Supply decisions also depend on the "shadow prices" of new discoveries and developed reserves, which are measures of the prices at which firms would be willing to buy or sell newly discovered sites and sites already prepared for extraction.[7] These shadow prices in turn depend on future prices of the resource and on exogenous factors that affect the value of reserves, such as an interruption of imports.

Finally, expectations of future profits depend on a host of institutional and regulatory factors, such as tax laws, price and production controls, public lands leasing practices, and the competitive structure of the market. Market structure is influenced in turn by the development of

7. Because these shadow prices are not always observable, in contrast to widely reported prices of extracted resources, they typically must be inferred from other available data. This issue is discussed further in chapters 4, 5, and 6; see also Tourinho (1979).

substitute goods, by changes in foreign supply conditions, and by government actions to control the market.

All of these factors, together with the basic assumption that firms act to maximize the expected net present value of their earnings over time, imply that nonrenewable resource supply is a dynamic process: current decisions are influenced by expectations about a variety of future events as well as by cumulative past activity. Moreover, the interconnections among stages in the supply process will be similarly dynamic. Finally, inherent uncertainties about the future affect current behavior. Consequently, a theory of supply is incomplete without a hypothesis about how expectations are formed and how firms respond to uncertainty. The dynamic and stochastic nature of the supply process, as indicated in the rest of this study, is the principal source of difficulty in drawing meaningful inferences and making reliable predictions about supply behavior.

On the practical side, the description of nonrenewable resource supply as a dynamic process indicates that static elasticity measures of supply responses are inadequate and misleading. The way in which a change in the price of a resource affects the rate of extraction and the rate of exploration cannot be described as a fixed relationship. Static elasticity relationships are inconsistent with the fundamental properties of nonrenewable resources, because depletion and changes in expectations will systematically alter these relationships over time.

It also follows that the *ceteris paribus* assumption commonly applied in static analysis is inappropriate in a dynamic setting. When measures of variables in the past, present, and future are necessarily linked to each other, a postulated change in the current value alone is inconsistent with the way the process works. For example, to understand how a change in current costs and prices affects current supply activities, one must take into account the effect of current supply activities on future costs and prices, because changes in future costs and prices also are affected by current decisions. Similarly, it is meaningless to posit a change in the current resource price without considering future prices, for it is the shape of the entire price path rather than a single point on the path that determines supply decisions. In addition, the influence of prices and costs at one stage of the supply process on the decisions made at other stages suggests that independent analyses of activities at a single stage are subject to error.

However, recognizing the dynamic complexities of the supply process, while conceptually desirable, is not easy to accomplish in practice. Dynamic models quickly become intractable or so complex that they offer few unambiguous guidelines for practical applications. These complexities are present even in the simplest characterization of the supply process, as demonstrated in chapters 2 and 3. Consequently, there is an absence of specificity in the theory about the form and content of decision rules, and this impedes empirical analysis by forcing analysts to adopt *ad hoc* specifications with questionable interpretations.

Uncertainty is particularly troublesome in modeling supply behavior, as indicated in chapter 4. Uncertainty about future prices and costs introduces conceptual questions about the manner in which producers respond to risk and the way they form expectations about future events. The common assumption is that entrepreneurs are risk neutral and seek to maximize the present value of expected net earnings. Alternatives to this "certainty equivalence" approach alter the analysis and the conclusions about supply behavior. The mechanism by which expectations are introduced also impinges on supply relationships and the stochastic properties of the model. In this book we focus on the "rational expectations" hypothesis as a promising approach to deal with these issues.

INDIVIDUAL VERSUS MARKET BEHAVIOR

Aggregation over individual decision units and individual resource deposits is necessitated by the type of data that are available and by the typical focus of interest on determinants of market behavior rather than individual behavior. The theory, however, concentrates on the analysis of decisions at the individual level, with relatively few extensions that describe market behavior. Consequently, empirical models of supply behavior are applications of the theory of individual decisions to industry data, or they are *ad hoc* specifications of industry behavior with no foundation in the theory of individual decisions.

Two primary considerations serve to differentiate market behavior from individual behavior in ways that defy description of the market by a "representative" firm in the industry. First, individual resource deposits differ at any given time in their geological and geophysical attributes, and the stock of deposits varies over time as a result of the cycle of depletion and new discoveries. These differences lead to firm-specific and site-specific characteristics of the cost of finding and extracting resources. Aggregations across these heterogeneous units will behave in ways that depend on the distribution of individual characteristics, yet these distributional factors have no relevance to individual decisions.

The second consideration is that market-determined prices and costs are independent of individual decisions but are variable and endogenously determined at the industry level of analysis. The dynamic interactions between supply and demand in various markets—for example, the markets for extracted output, new reserves, and factors of production—do not enter into the theory of individual decisions but are critical to the way the market behaves. Furthermore, these interactions are influenced by conditions that are incidental to individual decisions. The structure of the market—specifically, the degree of competition—is critical to specification of interactions among participants and to the concept of an equilibrium. Additionally, government intervention in the market in the form of price and production controls can alter the way the market performs beyond the implied changes in the behavior of individual firms.

Bridging the gap between analysis at the individual and the market levels is crucial to a practical understanding of the behavior of nonrenewable resource supply, yet the difficulties encountered in doing so are significant. Beyond the conceptual issues alluded to above, serious practical difficulties arise because of sparse data and because of inconsistencies between available data and the information requirements of the theory. Problems also arise in determining an estimation methodology that captures the dynamics of supply responses and identifies the separate influences in market data. These concerns are discussed in chapters 5 and 6, and recommendations for tackling the problems that remain are offered in chapter 7. Before turning to these market issues, however, we begin with a description of individual supply decisions.

2
The Basic Theory of Supply Behavior

The economic theory of nonrenewable resource supply discussed in this chapter is a description of how decisions concerning exploration, development, and extraction are made by individual profit-motivated firms. The core of the theory is a formal representation of depletion effects in the supply process and of how these effects influence the behavior of the firm.

Our purposes in this chapter are to describe the basic elements of the theory, to highlight their implications for the behavior of a firm, and to identify some of the obstacles to applying the theory. Of particular interest is the interdependence of decisions at all three stages of the supply process. Five simplifying assumptions will be maintained throughout this chapter: (1) there is no uncertainty—the firm has complete knowledge about current and future prices and costs; (2) there are no joint products—the firm produces a single homogeneous commodity (though it may operate many sites); (3) there are no cost "externalities" —the firm's costs depend only on its decisions and are independent of decisions by other firms; (4) the firm is a competitive price taker—it has no power to influence the price; and (5) the firm is fully integrated into all stages of the supply process.

The conceptual implications of relaxing these assumptions are explored in the following three chapters. Even at this simplest level of abstraction, however, the theory describes a complex dynamic process in which the nature of decision responses cannot be easily ascertained. In particular, the so-called r percent rule, which determines resource exploitation strictly by the rate of interest, is seen to be an oversimplified description of the supply process with little practical content. The discussion begins with the extraction decision with a fixed resource base, which is the context of the original Hotelling (1931) theory of supply. Subsequent sections introduce exploration for, and development of, new reserves and describe the behavioral implications of changes in price and cost conditions.

EXTRACTION FROM A FIXED RESERVE

The simplest version of resource supply theory concerns the decision about when and how fast to extract a resource from a fixed reserve of known dimensions. We first introduce notation and technical assumptions. The term q_t denotes the rate of extraction at time $t \geq 0$, and the stock of reserves available at time t is denoted by R_t. In the absence of exploration for and development of new reserves, reserves and extraction are related by the difference equation

$$R_{t+1} = R_t - q_t, \qquad t = 0, 1, \ldots \tag{1}$$

in which the initial reserves R_0 are given by some positive number $\overline{R}$.[1]

The cost of extraction is written $E(q, R)$, where the dependence on R indicates the effect of depletion on extraction cost. Specifically, it is assumed that extraction cost rises as the stock of available reserves declines. This assumption is denoted by $E_R(q, R) < 0$, where (throughout this study) variable subscripts indicate partial derivatives of functions with respect to the indicated argument. For an individual petroleum deposit, this relationship can be justified in physical terms by the decline in reservoir pressure and higher lifting cost as reserves decline. For hard-rock minerals and coal, a decline in reserves is a proxy for depletion of the thickest (or richest) seams and for the distance between the mine face and mine mouth.[2]

1. Hotelling's (1931) paper, and most of the extensions that followed it, are formulated in continuous time. A review of this literature can be found in Peterson and Fisher (1977). Following Kuller and Cummings (1974), Fisher (1982), and Hartwick (1978, 1982), our discussion is presented in discrete time. This simplifies the exposition and facilitates comparisons between theoretical and empirical specifications.

2. It is emphasized that reserves serve only as a first approximation of the depletion effect. For hard-rock minerals, in particular, a depletion-effects model should distinguish among grades of resources in the deposit, as developed by Farrow (1983) or Stollery (1983). In this context, the notion that the lowest cost grades or deposits are extracted first depends on the choice of numeraire (Kemp and Long, 1980b).

Marginal extraction cost is assumed to be positive (that is, $E_q > 0$) and, except where indicated, marginal extraction cost rises as the extraction rate increases for any level of reserves (that is, $E_{qq} > 0$). The latter assumption holds if there are "adjustment costs" to speeding up the rate of depletion by using less efficient means of extraction or by using existing capacity more intensively. Later, we consider the special case in which $E_{qq} = 0$ and $E_{qR} < 0$; that is, marginal extraction cost is constant with changes in output, yet increases as reserves decline. Finally, to simplify the discussion, the cost function is assumed to be unaffected by changes in extraction technology over time.[3]

The unit price of the extracted resource is denoted by p_t, and the price path $\{p_t, t \geq 0\}$ is known *ex ante* by the firm and is independent of the firm's extraction decision. The net present value of the firm's extraction plan over time $\{q_t, t \geq 0\}$ is given by

$$V(q_0, q_1, \ldots, q_T, T) = \sum_{t=0}^{T} d^t [p_t q_t - E(q_t, R_t)] \qquad (2)$$

where $d = 1/(1 + r)$ is the discount factor, $r > 0$ is the discount rate used by the firm in discounting net revenues back to the present, and $T \geq 0$ is the terminal date at which extraction from the deposit ceases. Note that the net present value depends on the choice of T as well as on the extraction plan $\{q_t\}$.

The firm is presumed to choose an extraction plan $\{q_t\}$ and a terminal date T that maximize equation (2) subject to the firm's knowledge of the price path $\{p_t\}$, the cost function $E(q, R)$, initial reserves $\overline{R}$, the stock change equation (1), the (obvious) nonnegativity condition $q_t \geq 0$, and the requirement that cumulative extraction cannot exceed the stock of initial reserves.[4] We do not rule out the possibility that extraction ceases before the stock of reserves is physically exhausted, in which case $R_{T+1} > 0$. This possibility emphasizes the distinction between physical exhaustion and economic exhaustion, where the latter occurs because the resource has no economic value, given current and future prices and costs.

The distinction between complete and incomplete physical exhaustion is important for theoretical and empirical reasons, as the discussion in this chapter is intended to make clear. The model of extraction developed in this section, like the extension to decisions about exploration and development later on, presumes that resources are not completely ex-

3. The effect of technical progress may be introduced by writing the cost function as $E(q, R, a)$, where a is a shift parameter reflecting the state of technical knowledge and E is negatively related to a (in other words, $E_a < 0$). Technical progress is discussed more fully in chapter 3.

4. For the time being, we ignore upper bounds on extraction from capacity constraints.

hausted. In contrast, the discussion of the "r percent rule" presumes complete exhaustion. As a practical matter, however, the assumption of incomplete physical exhaustion seems more appropriate. Resource deposits such as oil are not completely exhausted but are abandoned long before this point because they become uneconomic to extract. Other resources such as natural gas may exhibit total exhaustion of individual deposits, but because of high costs it is unlikely that all deposits ever will be discovered and developed. As long as any deposit remains untapped, incomplete physical exhaustion prevails. Other conceptual and empirical limitations of the complete exhaustion hypothesis and of the r percent rule are discussed below.

For theoretical completeness, we note that two conditions are sufficient (but not necessary) for incomplete exhaustion to emerge as a conclusion of the model being considered.[5] The first condition is that there exists a price $\bar{p} > 0$ where $p_t \leq \bar{p}$ for all t, so that the price path is uniformly bounded. Boundedness will occur if there is a finite price at which demand falls to zero. The second condition is that there is a stock level before complete exhaustion where the marginal cost of extraction exceeds the maximum possible price $\bar{p}$. In this case, each increment to extraction adds more to costs than to revenues as stocks decline to this critical level, eliminating the incentive to continue operation.

The Marginal Profit Rule

At each point in time, the choice of the extraction rate depends on the balance between marginal revenue—in this case, the price p_t—and the marginal opportunity cost of extraction. With incomplete exhaustion, the marginal opportunity cost has two components: the increase in current operating cost $E_q(q_t, R_t)$ from an increase in extraction and the increase in future extraction cost from the decrease in reserves. The second element is the depletion effect of a reduction in reserves, which in turn implies a reduction in the net present value of future earnings. At each time $s > t$ the opportunity cost of an increase in extraction at t is given by $-E_R(q_s, R_s) > 0$, indicating the rate of increase in cost with respect to the decline in reserves. For an extraction rate $q_t > 0$ to be optimal, given existing reserves of R_t, it is necessary that price equal the present value of these future costs plus the current marginal operating cost. Adding up the increases in future extraction cost and discounting them back to time t yields the equation

$$p_t = E_q(q_t, R_t) + \sum_{s=t+1}^{T} d^{s-t}[-E_R(q_s, R_s)] \tag{3}$$

5. See Toman (1982b) for a rigorous discussion of these conditions.

for a positive extraction rate at time t.[6] When the right-hand side of equation (3) exceeds p_t, extraction is not economical and $q_t = 0$ is the optimal rate. We return to this possibility below.

Equation (3) can be rewritten in equivalent form as

$$d^t\left[p_t - E_q(q_t, R_t)\right] = \sum_{s=t+1}^{T} d^s\left[-E_R(q_s, R_s)\right] \tag{4}$$

which states that discounted marginal profit on the last unit of resource extracted just equals the discounted sum of incremental increases in future extraction cost. The terms $\Sigma d^{s-t}(-E_R)$ and $\Sigma d^s(-E_R)$ in equations (3) and (4) are referred to as the "user cost" of extraction.[7]

Note that equations (3) and (4) are not defined for $t = T$. At the terminal date, the marginal profit condition reduces to

$$d^T\left[p_T - E_q(q_T, R_T)\right] = 0 \tag{5}$$

which states that discounted marginal extraction profit falls to zero when extraction ceases. This is a consequence of the assumption of incomplete exhaustion. As indicated in the next section, a different condition holds at the terminal date if the deposit is fully depleted.

Equations (3) through (5) have been presented as conditions for the optimal extraction path, though they may be interpreted as "stock value" or "portfolio balance" conditions for holding reserves as a durable asset in the ground.[8] To see this, let E_q^t and E_R^t denote values of the cost function derivatives at (q_t, R_t), subtract equation (3) at time $t+1$ from itself at time t, and use the equation $d = 1/(1+r)$ for the discount factor to obtain

$$\begin{aligned}\left(p_{t+1} - E_q^{t+1}\right) - \left(p_t - E_q^t\right) &= \sum_{s=t+2}^{T} d^{s-t-1}E_R^s - \sum_{s=t+1}^{T} d^{s-t}E_R^s \\ &= E_R^{t+1} + (1 - d^{-1}) \sum_{s=t+1}^{T} E_R^s \\ &= E_R^{t+1} + r\left(p_t - E_q^t\right)\end{aligned}$$

6. The discussion presumes that a program for maximizing net present value exists; see Toman (1982b) for a derivation of sufficient conditions for existence.

7. Because user cost equals the discounted marginal profit from extraction, it is also referred to in the literature as the net price, royalty, or rent. The difference between the user cost expressions in equations (3) and (4) is that cost increases are discounted back to time t in equation (3) but are discounted back to the initial date in equation (4).

8. See Solow and Wan (1976), Heal (1976), and Hartwick (1982) for analogous interpretations in the continuous time case.

Dividing both sides by $(p_t - E_q^t)$ and rearranging terms yields

$$\frac{\left(p_{t+1} - E_q^{t+1}\right) - \left(p_t - E_q^t\right)}{\left(p_t - E_q^t\right)} + \frac{-E_R^{t+1}}{\left(p_t - E_q^t\right)} = r \tag{6}$$

The first term on the left-hand side of equation (6) is the rate of capital gain from holding reserves in the ground rather than extracting them. The second term is the "dividend" rate, or the rate at which the "degradation charge" E_R^{t+1} is economized by holding reserves. Thus, if the subjective discount rate r also is assumed to be the market interest rate, then equation (6) implies that the net present value of extraction receipts is maximized when the rate of capital gain plus dividend on reserves in place equals the rate of interest. This is the equilibrium condition for holding reserves in the ground as an asset.

The marginal profit rule characterizes the optimal extraction path for any choice of the terminal date T. To determine the optimal terminal date as well as the extraction path, an additional condition is needed. Under incomplete exhaustion, the required condition is that discounted *average* profit and marginal profit fall to zero, so that the economic value of the deposit is exhausted (under prevailing prices and costs).

The necessity of this condition is illustrated in figures 2-1 and 2-2 for a finite choice of T.[9] In figure 2-1a the choice of T is too soon: price exceeds minimum average cost, so extraction can be profitably continued until cost rises high enough to choke off profit. In figure 2-1b extraction has continued too long: while price covers marginal cost, average cost has risen (due to depletion) beyond the point where extraction is profitable. Figure 2-2 shows the optimal configuration at the terminal time: price, marginal cost, and average cost all are equal. Since zero average profit implies zero total profit, the terminal condition can be formally written as[10]

$$d^T\left[p_T q_T - E\left(q_T, R_T\right)\right] = 0 \tag{7}$$

Equations (5) and (7) can be used to draw inferences about the magnitude of extraction as the terminal time is approached, and about the length of the operating horizon for different specifications of the extraction cost function. Suppose, first, that extraction is subject to fixed or "quasi-fixed" costs, where quasi-fixed costs do not vary with the rate

9. See Levhari and Liviatan (1977) for a formal proof. As indicated later in this chapter, an unbounded operating horizon ($T = \infty$) is a possible prediction of the model, even though the reserve stock is finite.

10. In formal terms, equation (7) is known as a "transversality condition." Another transversality condition is implicit in equations (4) and (5): as t approaches T, extraction user cost and therefore discounted extraction profit fall to zero.

of extraction but change over time as reserves are depleted.[11] In this case the average cost function is U-shaped, as shown in figure 2-2a, and the terminal output rate q_T is positive. Since initial reserves and cumulative extraction are finite, it follows that the terminal time must also be finite.[12] If there are no fixed or quasi-fixed extraction costs, marginal and average costs are equal only at $q=0$ and the optimal terminal configuration is illustrated in figure 2-2b, with q_t falling to zero as t approaches T. In this case the optimal terminal time may be finite or infinite.[13]

Implications for the Dynamics of Supply

Equation (3) shows that the extraction decision at time t depends on the current price p_t, on cumulative past extraction (through its effect on reserves), and on the initial size of the reserves (which determines the position of the marginal extraction cost function). Extraction also depends on user cost, and user cost depends on future extraction decisions. Future decisions in turn depend on the producer's expectations of future prices and costs as well as current conditions (since we have assumed no uncertainty, expectations coincide with actual outcomes). If the extraction cost function is shifted over time by technical progress and changes in input costs, then current decisions are also influenced by expectations about these changes.

These dynamic influences on extraction behavior characterize the resource supply function obtained by "solving" equations (3), (5), and (7), using a "backward recursion" argument.[14] To illustrate the argument, let T be any finite terminal time and, for the moment, treat R_T as fixed; then with $E_{qq}>0$, equation (5) can be inverted to yield

$$q_T = f_T(p_T, R_T) \tag{8}$$

The function f_T in equation (8) specifies the optimal rate of extraction in the final period, given the price p_T and remaining reserves R_T in that period.

11. An example could be the cost of maintaining fixed investments in extractive capacity.

12. If $T=\infty$ and $\lim_{t\to\infty} q_t > 0$, then $\sum_{t=0}^{\infty} q_t = \infty$. See Kemp and Long (1980a).

13. An example of an infinite horizon is given in Toman (1982b).

14. The following discussion is a somewhat heuristic application of the "dynamic programming" approach (Bellman, 1957; Bertsekas, 1976); see Burt and Cummings (1970) for an application to resource analysis. In practice, closed-form expressions for the supply functions cannot be derived except in certain special cases, and even in these cases the technical burden of the derivation can be substantial. The constant marginal cost model discussed in the next subsection is one example with a closed-form solution. Other examples based on linear-quadratic cost functions are presented in Epple and Hansen (1981) and Toman (1982b). Linear-quadratic models are discussed further in chapters 4 and 5.

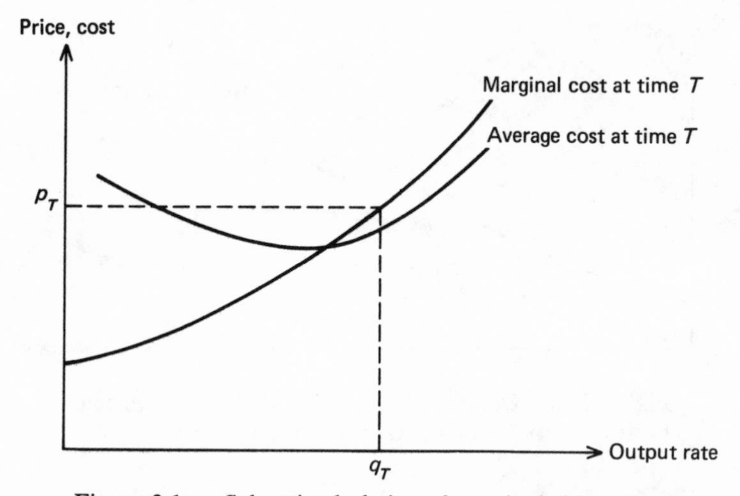

Figure 2-1a. Suboptimal choice of terminal time: T too small.

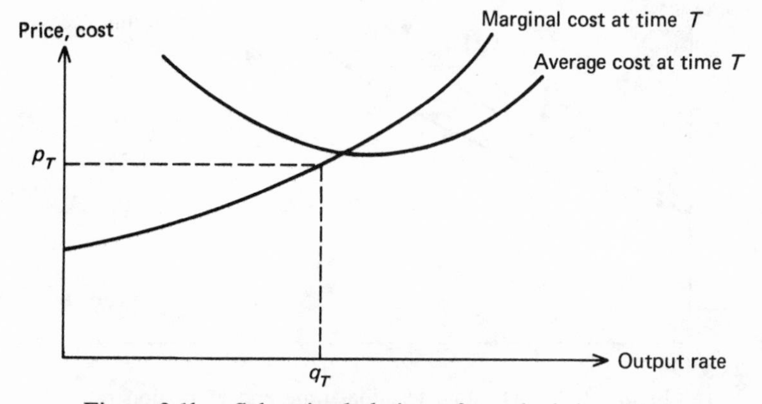

Figure 2-1b. Suboptimal choices of terminal time: T too large.

At time $T-1$ the marginal profit rule, equation (3), for extraction is

$$p_{T-1} = E_q(q_{T-1}, R_{T-1}) - dE_R(q_T, R_T) \tag{9}$$

Given that the period T decision is optimal, $f_T(p_T, R_T)$ from equation (8) can be substituted for q_T in equation (9). We also have $R_T = R_{T-1} - q_{T-1}$ from equation (1). Therefore, equation (9) can be written as

$$p_{T-1} = E_q(q_{T-1}, R_{T-1}) - dE_R[f_T(p_T, R_{T-1} - q_{T-1}), R_{T-1} - q_{T-1}] \tag{10}$$

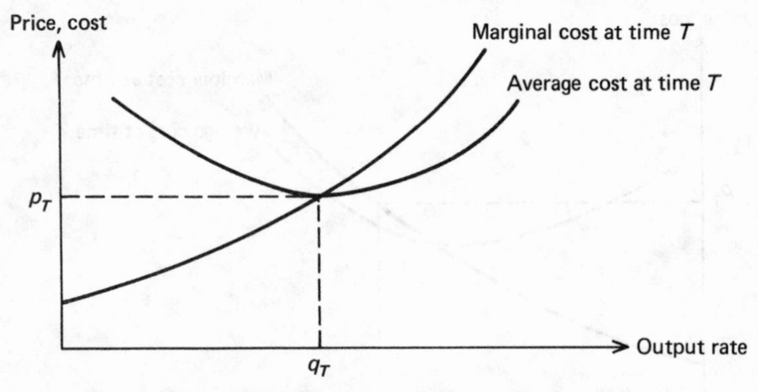

Figure 2-2a. Optimal terminal time and extraction rate: Positive quasi-fixed cost; U-shaped average cost.

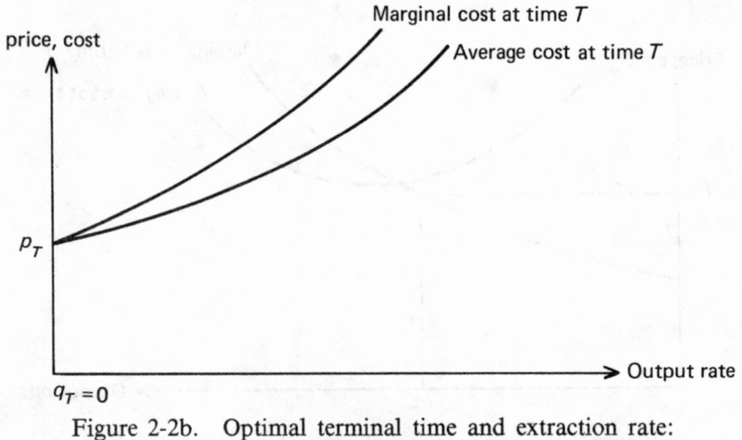

Figure 2-2b. Optimal terminal time and extraction rate: Zero quasi-fixed cost; increasing average cost.

With R_{T-1} fixed, equation (10) can be solved for q_{T-1} to yield

$$q_{T-1} = f_{T-1}(p_{T-1}, p_T, R_{T-1}) \tag{11}$$

provided the extraction cost function is suitably well behaved.[15]

Proceeding in this fashion, extracted resource supply functions

$$q_t = f_t(p_t, p_{t+1}, \ldots, p_T, R_t) \tag{12}$$

15. The required condition is that $E(q, R)$ is strictly convex in q and R, so that $E_{qq} > 0$, $E_{RR} > 0$, and $E_{qq}E_{RR} - E_{qR}E_{Rq} > 0$.

can be derived for each $t = 0, 1, \ldots, T$; these equations subsume equations (8) and (11) as special cases. For each t, f_t specifies the extraction rate that maximizes net present value given R_t reserves remaining for extraction and the price path $\{p_t, p_{t+1}, \ldots, p_T\}$. The actual path of extraction rates for an arbitrary choice of T is determined by making use of equation (1) and the initial condition $R_0 = \overline{R}$: $q_0 = f_0(p_0, \ldots, p_T, \overline{R})$, $q_1 = f_1[p_1, \ldots, p_T, \overline{R}_0 - f_0(p_0, \ldots, p_T, \overline{R})]$, and so forth.

Finally, the optimal choice of terminal time can be determined by following the solution path until equation (7) holds; that is, until cost rises to the point where profitable extraction opportunities are eliminated. As noted above, the terminal date may be infinite. In this case, the solution can be derived (at least heuristically) from the "limit" of the above argument as the terminal date grows large. The extraction supply function in this case will have the general form

$$q_t = f_t(p_t, p_{t+1}, \ldots, R_t) \tag{13}$$

Note that if the cost function included an exogenous shift parameter to account for technical progress or input prices, as in $C(q, R, a)$, the supply function would also depend on current and (expected) future values of these parameters:

$$q_t = \begin{cases} f_t(p_t, \ldots, p_T, a_t, \ldots, a_T, R_t), & T < \infty \\ f_t(p_t, p_{t+1}, \ldots, a_t, a_{t+1}, \ldots, R_t), & T = \infty \end{cases} \tag{14}$$

This general line of argument can also be applied to the more complex models of reserve discovery and development discussed below.

Constant Marginal Extraction Cost

For many nonrenewable resources, such as petroleum, it may be reasonable to assume that marginal direct operating cost increases only slightly with the extraction rate until the limit of extractive capacity is approached. In the extreme case of marginal extraction cost that is independent of output but rises as reserves decline, supply responses will have different characteristics than those represented by the general equations (12) through (14). Specifically, optimal extraction is equal to the capacity rate if price exceeds marginal operating cost plus user cost, and equal to zero if price is below cost. Only if price and marginal opportunity cost happen to be equal will extraction be at some rate between zero and capacity.

To investigate extraction behavior with constant marginal cost, we adapt a continuous time model developed by Pindyck (1981a) with an exponential growth path $p_t = p_0 e^{\alpha t}$ for price (so that the price grows over time at a constant rate α). The model provides a simple yet informative characterization of supply behavior in which supply functions can be calculated explicitly. Note first that $\alpha < r$ is required for extraction to occur, for if the price grows at the same rate as the discount rate or faster, then reserves in the ground are appreciating at least as fast as their extracted value and thus will not be extracted. The cost of extraction is given by $E(q, R) = qh(R)$, with unit cost inversely related to reserves so that $h' < 0$. Since $E_{qq} = 0$ and $E_{qR} = h'(R) < 0$ (that is, marginal cost is constant with respect to extraction and rises as reserves decline), it can be shown that the optimal extraction path satisfies

$$q_t = \begin{cases} \bar{q}, & p - E_q > U \\ q_t^*, & p - E_q = U \\ 0, & p - E_q < U \end{cases} \tag{15}$$

In this equation, $\bar{q}$ is a fixed limit on extractive capacity, U is user cost at time t [the analogue of the sum on the right side of equation (3)], $p - E_q = p - h(R)$, and q_t^* is a "singular path" where price equals marginal operating plus user cost.[16] This path is found by differentiating the marginal profit rule $p - E_q = U$ with respect to time and is implicitly given by

$$h(R_t^*) = \frac{r - \alpha}{r} p_t \tag{16}$$

where R_t^* is the reserve stock along the singular path. Equation (16) states that along the singular path, marginal extraction cost is proportional (but not equal) to price. Moreover, since user cost equals the difference between price and marginal cost $h(R)$, it follows from equation (16) that

$$U_t = \frac{\alpha}{r} p_t \tag{17}$$

so that user cost also is proportional to price along the singular path. Differentiating both sides of equation (16) with respect to time, and using the equations $\dot{q} = -R$ and $\dot{p} = \alpha p$, it follows that the singular path satisfies

$$q_t^* = \frac{(r - \alpha)\alpha p_t}{-rh'(R_t^*)} > 0 \tag{18}$$

Which of the three alternatives in equation (15) prevails depends on the price path and the magnitude of marginal extraction cost. If, for example, the initial price is less than marginal cost, then $q_t = 0$ until price rises to the point of covering both operating cost and user cost. On the other hand, if initial reserves are large and marginal user cost is small, output will be at capacity until reserves are partially depleted and user

16. It is implicitly assumed that $0 \le q_t^* \le \bar{q}$.

costs increase. Similarly, if the price is constant, discounted revenues decline over time and there is no advantage to deferring output. Extraction will be set at capacity until marginal cost rises to the level of price, at which point the resource deposit will be abandoned.

To investigate the linkages among the three paths, consider the special case of unit extraction cost that is inversely proportional to reserves, so that $h(R)=m/R$ and $h'(R)= -m/R^2$ for some $m>0$. Substituting these expressions and the price formula $p_t=p_0e^{\alpha t}$ into equations (16) and (18), we obtain the expressions

$$R_t^* = \frac{mr}{(r-\alpha)p_0}e^{-\alpha t}$$
$$q_t^* = \frac{\alpha mr}{(r-\alpha)p_0}e^{-\alpha t} \tag{19}$$

for reserves and output along the singular path.[17] The behavior of extraction over the entire operating horizon then can be described graphically as shown in figure 2-3.[18] If the initial reserve stock $\overline{R}$ is large, as in figure 2-3a, so that unit cost $m/\overline{R}$ is low relative to the initial price p_0, then output is initially set at capacity until reserves $R_t=\overline{R}-t\overline{q}$ fall to the singular path level R_t^*. At this time (denoted τ in the diagram), output jumps down to the singular path q_t^* and follows this path thereafter. If initial reserves are low, so that unit cost is high, then the opposite pattern emerges as shown in figure 2-3b: output is initially zero until price rises, then jumps up to the singular path.

An interesting feature of this optimal output path is that along the singular trajectory given by equation (19) both output and reserves decline exponentially at a rate equal to the price growth rate, and the reserves-to-production ratio R^*/q^* is constant and equal to the reciprocal of the price growth rate. Thus, as noted by Pindyck, the assumptions of a unit cost inversely proportional to reserves and of constant-percentage price growth provide a theoretical rationale for the "decline curve" hypothesis of exponential output decline with a constant R/q ratio that is commonly used to describe supply behavior in the petroleum industry.[19]

17. Pindyck's formulas for R^* and q^* (1981a, p. 281), which differ from equation (19), are incorrect, but the error does not affect his conclusions. Note from equation (19) that $q_t^*=\alpha mr/(r-\alpha)p_t$; while q_t^* appears to depend only on the current price p_t, it also depends implicitly on future prices through its dependence on the price growth rate α. Note also that while q_t^* is *inversely* related to the current price p_t, it is directly related to the price growth rate: an increase in α increases q_t^*.

18. These diagrams are adapted from Pindyck (1981a, p. 282).

19. See, for example, Bradley (1967) and Adelman (1972) for discussion of decline curves. Also, refer to the discussion of optimization models in chapter 6.

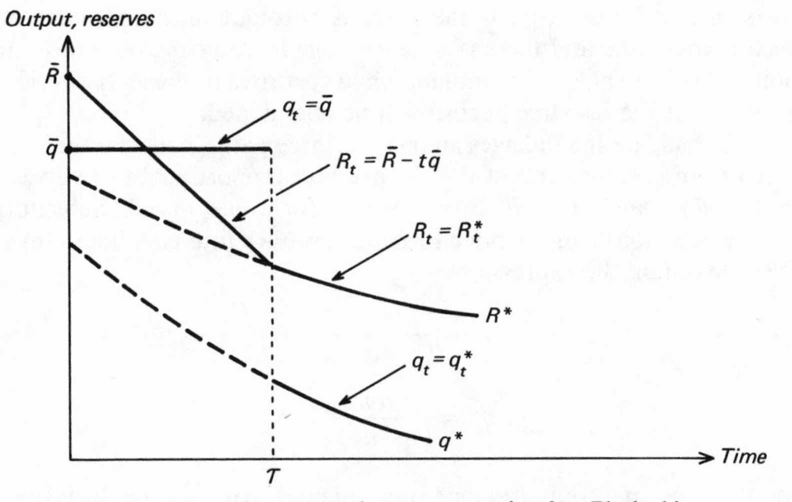

Figure 2-3a. Output and reserves paths for Pindyck's (1981a) constant marginal cost model: Large initial reserves.

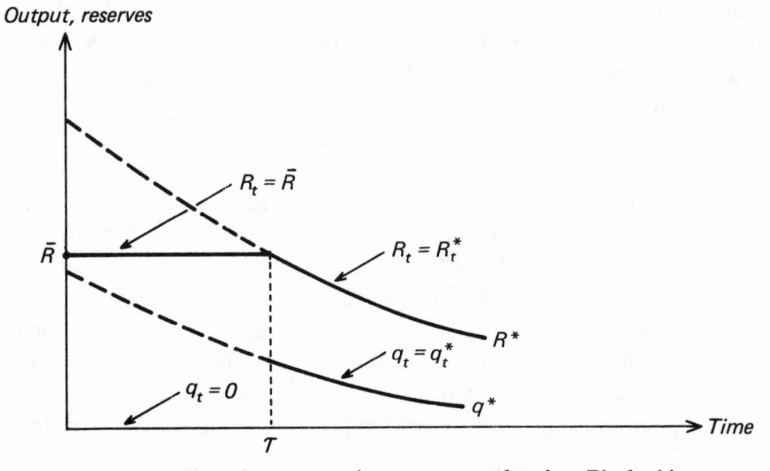

Figure 2-3b. Output and reserves paths for Pindyck's (1981a) constant marginal cost model: Small initial reserves.

The model also draws attention to the importance of extraction capacity in determining extraction behavior. In particular, figure 2-3a shows extraction equal to capacity during the initial phase of operations; moreover, an increase in the capacity rate through development investment raises the initial extraction rate while speeding depletion and moving up the time τ in figure 2-3 when output switches to the decline curve. This pattern of behavior depends on the assumptions of a constant capacity rate and a marginal extraction cost that is independent of the output rate. Subsequent parts of this chapter discuss in greater detail the implications of capacity rates and the sensitivity of output paths to assumptions about cost and capacity.

THE *r* PERCENT RULE

Many studies of nonrenewable resource production and pricing have relied on a simplified version of the Hotelling model in which extraction cost does not vary with cumulative past output or remaining reserves.[20] There are, however, both theoretical and practical reasons to doubt the applicability of this characterization of supply behavior.

To begin with, if extraction cost does not vary with depletion, complete exhaustion of the resource stock is inevitable unless an unexpected event —such as the development of a lower cost substitute—permanently drives the resource price below the minimum unit cost of extraction. Unlike the model discussed above, declining reserves will not increase cost to choke off extraction before depletion. The marginal profit rule for maximizing net present value also is simpler than that given by equation (4). In the present case, this rule, known as the "*r* percent rule" or "Hotelling law," states that the extraction path must be chosen so that discounted marginal profit is constant over time, so that undiscounted marginal profit grows at the discount rate. In symbols,

$$d^t\left[p_t - E_q(q_t)\right] = \lambda \tag{20}$$

where E_q is the marginal cost of extraction and $\lambda > 0$ is a constant that represents a pure "scarcity rent" associated with the finiteness of the resource stock.

To understand the reasoning behind equation (20), consider the costs and benefits of extracting an additional unit of resource at any date t.[21] The increase in revenue is the price p_t, while the increase in current cost is the marginal cost of extraction E_q^t. Additional extraction at time t also involves a user cost, but unlike the model discussed in the previous

20. Examples can be found in Dasgupta and Heal (1979) and the references in Devarajan and Fisher (1981).

21. For a more extensive discussion, see Herfindahl and Kneese (1974).

section, this user cost does not reflect "degradation" of the resource and consequent increases in future extraction cost—the sum on the right-hand side of equation (3) or (4). Instead, the user cost consists solely of the value of extraction forgone at the terminal time T. This opportunity cost arises because, holding other decisions constant, increased current extraction necessarily decreases terminal extraction under complete exhaustion. The present value of this future revenue loss is $d^T[p_T - E_q^T]$, and the present value of the current revenue gain is $d^t[p_t - E_q^t]$. To maximize net present value, these two terms must be equal; if they are not equal, then net present value can be increased by shifting output between dates t and T until they are. Letting $\lambda = d^T[p_T - E_q^T] > 0$ denote the user cost of extraction, we arrive at equation (20).

Like equation (6), equation (20) can be interpreted as an equilibrium condition on the asset value of unextracted reserves.[22] As in the interpretation of equation (6), marginal profit also is the shadow price of an additional unit of reserves. With $E_R = 0$, so that there is no degradation charge to extraction, then the only return to holding reserves (rather than extracting them) is the capital gain. For reserves to be an investment worth holding, the rate of capital gain—the rate of increase in marginal profit—must equal the discount rate, which is the rate of return on other investments.

Inferences about the path of nonrenewable resource prices over time can be deduced from equation (20) by rearranging terms:

$$p_t = E_q^t + d^{-t}\lambda, \qquad t = 0, 1, \ldots. \tag{21}$$

Using equation (21), together with the fact that λ is constant and $d = 1/(1 + r)$, it can be shown that

$$\Delta p_t/p_t = r + \Delta E_q^t/p_t - rE_q^t/p_t \tag{22}$$

where $\Delta p_t = p_{t+1} - p_t$ is the change in price and $\Delta E_q^t = E_q^{t+1} - E_q^t$ is the change in marginal extraction cost. If marginal extraction cost is constant over time, so that $\Delta E_q^t = 0$, then equation (22) implies that the percentage growth in price is equal to the discount rate less a term that is inversely proportional to the resource price level. As the magnitude of p_t increases, this second term diminishes, and the rate of price growth approaches the discount rate.[23]

The r percent rule provides a simple and internally consistent characterization of nonrenewable resource supply, particularly its interpre-

22. See Solow (1974) for an extensive discussion of this interpretation. Note that if $E_R = 0$ in equation (6), then marginal extraction profit should grow over time at the same rate as the discount rate and the result is the same as is given by equation (20).

23. Note that as E_q^t approaches zero and ΔE_q^t approaches zero in equation (22), the rate of price growth approaches the discount rate at all dates.

tation of the necessary rate of return on nonrenewable resources relative to other assets. However, the stringent assumptions upon which the rule is based raise grave doubts about its theoretical and practical applicability. Note first that the derivation of equation (20) ignores both depletion effects (rising costs at individual sites and exploitation of lower grade deposits) and the discovery and development of new reserves or substitute goods. As indicated in the discussion of equation (6), these factors lead to incomplete exhaustion and to a rate of increase in marginal profit that is less than the discount rate. In addition, these factors lead to cost changes that push resource prices off the path of almost exponential growth predicted by equation (22). Uncertainties about future prices and costs, producers' attitudes toward risk, technical progress, and changing market structures and regulatory environments also weaken the predictive power of the *r* percent rule.

Empirical support for the rule also is scant. In a detailed econometric analysis of resource prices, Slade (1982a, 1982b) finds substantial indirect evidence that resource price trends have been influenced by both depletion effects and technical change. In direct tests of the rule by Smith (1981), variants of equation (22) are fitted to resource price data with little success.

These observations suggest that the *r* percent rule has limited practical applicability, despite its widespread use. To fully understand the dynamics of resource supply, it is necessary to use more complex models that recognize both depletion effects and additions to reserves through exploration and development.[24] With regard to the interpretation of resource stocks as an asset, note that the value of reserves is determined not by physical size alone but by their contribution to lower extraction cost and increased extractive capacity. Viewed in this way, the value of reserves changes over time as old reserves are depleted and new reserves are added. Consequently, there is no reason why rates of return *per physical unit* of the resource (that is, the rate of return per barrel or ton of reserves) should be equal to the discount rate.[25] This topic arises again in the next section.

24. A partial exception to this statement is noted by Hartwick (1982) in the case in which there is a finite number of resource grades and marginal extraction cost is a constant for each grade. During the time interval for which each grade of resource is extracted, marginal profit will rise at the rate of interest. But when the transition is made to the next more costly grade, the rate of marginal profit falls abruptly before resuming the exponential growth path.

25. Changes in interest rates are another potential source of divergence between rates of return on resource stocks and rates of return on other assets. Resource stocks are durable goods whose size and composition cannot instantly respond to such changes. The implicit assumption of a single, constant discount rate in the marginal profit rules (3) and (4) is only for ease of exposition.

DEVELOPMENT AND DISCOVERY OF NEW RESERVES

A model of the extraction decision alone is an incomplete description of supply activity because it presupposes a fixed stock of existing reserves, whereas the reserve stock can change with new additions as a result of exploration and development activities. The supply process involves decisions about extraction and reserve additions, with the two decisions being jointly determined by factors influencing the stream of net revenues over time. Decisions about new reserves can substantially alter the optimal extraction path.

We begin by examining a model similar to one presented by Peterson (1978) for extraction and development decisions by an individual firm. With the extension of the model into other stages of the supply process, additional notation and assumptions are required. Let y denote the gross rate of new additions to developed reserves, and let z denote the cumulative volume of past reserve additions. For convenience, assume that no prior development has taken place, so that $z_0 = 0$. When $z_0 > 0$ the analysis in this section still applies, with z_t redefined to be cumulative additions since the initial time $t = 0$.

Because the firm can add to reserves and can decrease them through extraction, the reserves equation (1) must be amended to

$$R_{t+1} = R_t - q_t + y_t, \qquad t = 0, 1, \ldots; \qquad R_0 = \overline{R} \tag{23}$$

Cumulative additions follow the difference equation

$$z_{t+1} = z_t + y_t, \qquad t = 0, 1, \ldots; \qquad z_0 = 0 \tag{24}$$

The cost of adding new reserves is given by a function $C(y, z)$, which depends on both the rate of new additions and cumulative past additions. The dependence of development cost on z reflects the impact of depletion on supply. Assume that $C_z > 0$, indicating rising development cost as cumulative past additions grow and reserves remaining to be developed decline. This cost specification follows if the lowest cost deposits are the first to be developed and extracted, implying one-to-one correspondence between increases in cumulative additions, declines in deposit quality, and upward shifts in development cost over time. However, this specification of depletion effects in terms of cumulative additions is not always valid, as noted earlier, and should be regarded as a first approximation to a potentially more complex process. Also assume, for the moment, that there is no distinction between development activity on the intensive and the extensive margins. Both assumptions are examined more closely below.

The objective of the integrated mining firm is to maximize the present value of net revenues, given by

$$V(q_0, \ldots, q_T, y_0, \ldots, y_T, T) = \sum_{t=0}^{T} d^t [p_t q_t - E(q_t, R_t) - C(y_t, z_t)] \quad (25)$$

subject to the dynamic equations (23) and (24) for developed reserves and cumulative additions and the obvious requirement that both developed reserves and the decision variables be nonnegative. Cumulative reserve additions also are subject to the theoretical limit $z_T \leq \bar{z}$, where $\bar{z}$ is the physical stock of reserves remaining to be developed at the initial time $t = 0$ (that is, the undeveloped "resource base"). However, we assume throughout this section that depletion effects choke off both extraction and development before complete physical exhaustion occurs.[26]

Marginal Profit Rules for Extraction and Development

The general form of the marginal profit rule for extraction does not change when development of new reserves enters the picture. A balance is struck between the price of the extracted resource and the opportunity cost of extraction, including user cost, and conditions (3) through (5) remain valid. What does change, however, is the content of the marginal profit rule for extraction, because the user cost term on the right-hand side of equation (3) depends on future levels of reserves, and reserves in turn depend on the path of development decisions.

To complete the picture, a marginal profit rule for reserve additions is required. As with extraction, the marginal opportunity cost of adding new reserves is the sum of current marginal development cost C_y and a user cost. User cost is the discounted sum of future increases in development cost, denoted by C_z, which result from increased reserve additions that drive up cumulative additions and drive down the remaining inventory of undeveloped reserves. These costs are the "degradation charges" associated with current additions to reserves, and are analogous to the degradation charges $-E_R$ in the user cost of extraction.

To determine the optimal rate of addition to reserves, the marginal opportunity cost must be balanced against the benefit (in present value terms) from adding new reserves. The benefit, or shadow price, of new reserves is precisely the user cost of extraction. To see this, note that the effect on future developed reserves and extraction costs of an incremental addition to reserves exactly offsets the effect of an equal increment to current extraction. Thus, the incremental value of new reserve additions

26. For a precise formulation, see Toman (1982b).

is just the present value of future extraction cost savings from growth in the reserve stock, and the latter is the user cost of extraction.

The balance between the costs and benefits of reserve additions can be expressed in formal terms by the following marginal profit rule for development decisions:

$$C(y_t, z_t) + \sum_{s=t+1}^{T} d^{s-t} C_z(y_s, z_s) \geq \sum_{s=t+1}^{T} d^{s-t}[-E_R(q_s, R_s)],$$

$$y_t\left[C_y(y_t, z_t) + \sum_{s=t+1}^{T} d^{s-t} C_z(y_s, z_s) - \sum_{s=t+1}^{T} d^{s-t}[-E_R(q_s, R_s)]\right] = 0 \tag{26}$$

for $t = 0, \ldots, T-1$, where the sum $\Sigma d^{s-t} C_z$ is the user cost of reserve additions and $\Sigma d^{s-t}[-E_R]$ is the shadow price of new reserves. At the terminal time T, both development user cost and the shadow price of new reserves are zero, since by definition there are neither future costs nor future benefits to adding reserves at the terminal date. Thus, for $t = T$ the marginal profit rule reduces to

$$C_y(y_T, z_T) \geq 0,$$
$$y_T C_y(y_T, z_T) = 0 \tag{27}$$

Expressions (26) and (27) are stated as inequalities to emphasize the possibility that development activity may cease before the terminal time T for extraction. Note first that $y_T = 0$ by equation (27), indicating that it can never be optimal to add to reserves at the terminal date when they will not be used. Moreover, if $C_y(0, z) > 0$, meaning that marginal development cost is strictly greater than zero even as reserve additions fall to zero, then reserve additions will cease before the terminal time since their shadow price declines to zero as t approaches T. Thus, the solution path predicted in this case generally conforms to observed behavior for a single deposit: an initial phase of contemporaneous development and extraction, followed by a terminal phase in which reserves are depleted to their economic limit. If $C_y(0, z) = 0$, which is less plausible, then new reserves may be added right up to the terminal date T.[27] These alternatives are contrasted in figure 2-4.

Equations (3), (5), (26), and (27) describe the optimal paths of extraction and development for any choice of the terminal date T, given the resource price path $\{p_t, t \geq 0\}$ and the initial reserve stock $\bar{R}$. To round

27. See Pindyck (1978c) for further discussion of this point. Pindyck also notes that if initial reserves are large, it may be optimal to defer additional development until reserves have been somewhat depleted. This possibility is allowed for by expressions (26) and (27).

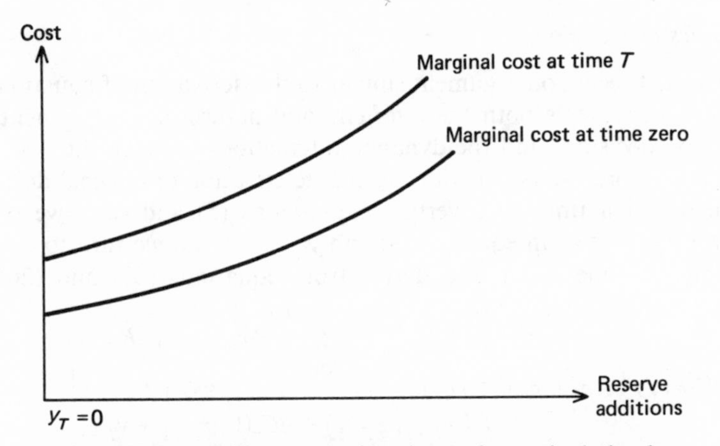

Figure 2-4a. Different specifications of marginal development cost: Reserve additions fall to zero before abandonment.

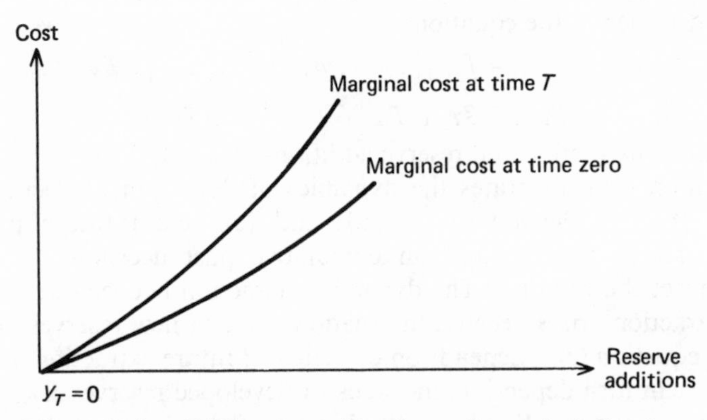

Figure 2-4b. Different specifications of marginal development cost: Reserve additions fall to zero only at abandonment.

out the solution, the additional condition of zero terminal profit

$$d^T[p_T q_T - E(q_T, R_T) - C(y_T, z_T)] = 0 \tag{28}$$

is required for reasons indicated above in connection with extraction. In principle, these five equations can be solved to determine the extraction and development paths, the terminal time T, and (implicitly) the time $T' \leq T$ when reserve additions cease. We turn next to some important properties of this solution.

Dynamics of Decisions

A backward recursion argument similar to the derivation of equations (8) through (12) reveals both the intertemporal nature of development and extraction decisions and the dynamic interactions between the two decisions. As before, we begin with the characterization of optimal decisions at the terminal time T. Inverting expressions (5) and (27), we obtain $q_T = f_T(p_T, R_T)$ as in equation (8) and $y_T = 0$. Next, we turn to optimal decisions at time $T-1$ and derive from equations (23) and (26) the following equations:

$$P_{T-1} - E_q(q_{T-1}, R_{T-1}) + dE_R[f_T(p_T, R_{T-1} - q_{T-1} + y_{T-1}), R_{T-1} - q_{T-1} + y_{T-1}] = 0 \quad (29)$$

$$C(y_{T-1}, z_{T-1}) + dC_z(0, z_{T-1} + y_{T-1}) + dE_R[f_T(p_T, R_{T-1} - q_{T-1} + y_{T-1}), R_{T-1} - q_{T-1} + y_{T-1}] = 0 \quad (30)$$

Equation (29) is analogous to equation (10). In equation (30), $C_z(0, z_{T-1} + y_{T-1}) = C_z(0, z_T)$ is the rate at which reserve additions at time $T-1$ increase quasi-fixed development cost $C(0, z_T)$ at time T. Provided the cost functions satisfy certain conditions, equations (29) and (30) can be inverted to derive the equations

$$q_{T-1} = f_{T-1}(p_{T-1}, p_T, R_{T-1}, z_{T-1}) \quad (31)$$

$$y_{T-1} = g_{T-1}(p_{T-1}, p_T, R_{T-1}, z_{T-1}) \quad (32)$$

for optimal extraction and reserve additions at $T-1$.[28]

Equation (32) illustrates the dynamics of development decisions: reserve additions depend on current and (expected) future prices of the extracted resource and on cumulative past decisions to develop *and* extract the resource. The dynamic interaction between development and extraction arises because the shadow price of new reserves, the term dE_R in equation (30), depends on current and future extraction decisions, and these in turn depend on the stock of developed reserves R_{T-1} as well as on p_{T-1} and p_T. Similarly, the impact of development decisions on extraction decisions is indicated in equation (31) by the dependence of q_{T-1} on cumulative reserves additions z_{T-1} as well as on developed reserves R_{T-1}. This influence arises because the user cost of extraction depends on the planned levels of reserves in subsequent periods, and planned reserves depend on planned development decisions.

The same interactions appear in the equations

$$q_t = f_t(p_t, p_{t+1}, \ldots p_T, R_t, z_t) \quad (33)$$

$$y_t = g_t(p_t, p_{t+1}, \ldots p_T, R_t, z_t) \quad (34)$$

28. These solutions are well defined if the extraction and development cost functions are strictly convex in both their arguments.

for extraction and additions to reserves at all dates $t < T$. Extraction decisions are independent of changes in cumulative reserve additions only if development decisions are independent of z_t. This condition holds only if there are no depletion effects at the development stage, so that $C = C(y)$ and $C_z = 0$. Similarly, decisions at both stages are independent of R_t only if there are no depletion effects from extraction, so that $E = E(q)$ and $E_R = 0$.

The dynamic interactions among stages in the supply process have important practical implications for empirical analysis of resource supply behavior. It is clear that empirical analysis of individual stages of the process will not capture these interactions and are subject to serious error. The behavior of extracted output depends on the behavior of reserve additions, and vice versa, so that simultaneous analysis of all stages in the supply process is required to assess these influences. This point is elaborated upon in the next section and in chapters 5 and 6.

Capacity Constraints on Extraction

The decision rule for extraction described above presumes that the optimal rate of extraction is not constrained by productive capacity of the resource deposit. When capacity is binding, the optimal extraction path can deviate substantially from the unconstrained extraction path, as noted earlier in the section on constant marginal extraction cost. Thus, for completeness, it is necessary to examine the case in which capacity is a binding constraint. If capacity constraints are binding, the general form of the marginal profit rules discussed above remains unchanged. However, the extraction user cost term $\Sigma d^{s-t}(-E_R)$ must be revised because the presence of binding capacity constraints imposes an additional opportunity cost on extraction: if extraction reduces capacity, an incremental increase in current output reduces the present value of future capacity.

To determine the effect on the extraction decision, let p_t^R be the user cost of extraction discounted back to date t. Assume that extractive capacity can be represented by a function $F(R)$ of the developed reserve stock. Let F' denote the derivative of F, and assume that $F' > 0$, so that an increase in reserves also increases capacity. Now consider how an increase in extraction time at t will affect future extraction cost and capacity. As before, one component of user cost is the present value of future extraction cost increases, regardless of whether capacity is binding in the future. The second component, when capacity is binding at a future date s $(s > t)$, is equal to the change in capacity, $F'(R_s)$, times the net present value of lost output. The net present value of lost output is measured by the marginal profit of extracted output at time s, given by

$p_s - E_q^s$, less the reduction in user cost for periods beyond time s because output is lower, given by p_s^R.

Putting these pieces together, we obtain the user cost formula

$$p_t^R = \sum_{s=t+1}^{T} d^{s-t}[-E_R(q_s, R_s)] + \sum_{s \in J_t} d^{s-t} F'(R_s)\left[p_s - E_q(q_s, R_s) - p_s^R\right] \tag{35}$$

where J_t is the set of all dates $s > t$ at which the capacity constraint is binding. If capacity is binding at every time $s > t$, the second sum is taken over all $s = t+1, \ldots, T$. If capacity is never binding, the second sum vanishes and the user cost reduces to the sum in equation (3). Note also that if capacity is independent of reserves, so that $F'(R_s) = 0$, then the second sum in equation (35) also vanishes.

The marginal profit rule for reserve additions expressed in equations (26) and (27) continues to hold with p_t^R in place of $\Sigma d^{s-t}(-E_R)$ as the shadow price of new reserves. With the same substitution, the extraction marginal profit rule given by equations (3) and (5) holds in all periods when capacity is nonbinding, while $q_t = F(R_t)$ when capacity is binding. Again, these equations plus the zero terminal profit condition (28) can be solved, in principle, to derive extraction and reserve decision functions like equations (33) and (34), with decisions depending on current and future prices, developed reserves, and cumulative past reserve additions. However, the properties of these functions—in particular, how output and reserve additions respond to depletion and changes in price—will also depend on the characteristics of the capacity constraint function and the extent to which capacity is binding.

This discussion raises the practical question of how to infer the effect (if any) of potential capacity constraints on supply decisions. If separate data on capacity and output are available, the problem is not serious because the capacity constraint can be estimated. As noted in chapter 6, however, data on crude oil and natural gas extractive capacity are scarce and unreliable, making it difficult to distinguish the effects of these constraints from other influences on behavior.

To illustrate the practical problems, consider a slight modification of the constant marginal extraction cost model discussed earlier, where the price is assumed to rise at a constant rate α, and extraction declines at the rate α after an initial phase of capacity output at a constant rate $\bar{q}$ (see equation (19) and figure 2-3a). Suppose that capacity is proportional to reserves instead of being constant, so that $F(R) = \beta R$ with $\beta > 0$ a proportionality constant. If output is set equal to capacity in every period and no new reserves are added, then extraction and reserves will follow the paths

$$\overline{R}_t = \overline{R}e^{-\beta t}; \quad \bar{q}_t = \beta \overline{R}e^{-\beta t} = \beta \overline{R}_t; \quad \overline{R}_t/\bar{q}_t = 1/\beta \tag{36}$$

With initial reserves that are large and $\beta > \alpha$, the optimal paths of output, capacity, and reserves given by equations (19) and (36) are shown in figure 2-5.[29] There is an initial phase of capacity extraction determined by equation (36) until $\overline{R}_t$ falls to the singular path R_t^* determined by equation (19). At this point, output falls to the singular path q_t^* and reserves decline at a slower rate along R_t^*. Because extraction is below capacity in the second phase, capacity βR_t^* lies above the path $\bar{q}_t = \beta \overline{R}_t$. Note that during each phase, output and reserves decline at a constant rate (β in the first phase, α in the second) and the reserves-to-production ratio is constant, though it is lower in the capacity-constrained phase than in the unconstrained phase.[30]

Under the assumptions of the model, the effect of the extraction capacity constraint on a single resource deposit can be unambiguously determined by comparing reserves-to-production ratios across time. If all individual deposits had identical initial reserve levels, marginal extraction costs, and capacity decline rates, the same determination would be possible using aggregate data for all deposits and firms. In practice, however, deposits vary in the level of reserves and in the geophysical characteristics that determine cost and capacity, as discussed in chapters 5 and 6. Therefore, at any moment in time, some fraction of "younger" deposits will be operating on their capacity decline curves while the remaining "older" deposits will be operating in the terminal decline curve stage. The individual decline curves will vary among deposits within each phase and the mix of old and young fields will change over time. Consequently, to assess the effect of capacity limitations on aggregate supply requires knowledge of the distribution of deposits by age and geophysical characteristic, and knowledge of how these distributions change over time.[31]

Divergence between individual firm and aggregate supply behavior is a recurrent and troublesome issue that goes well beyond the question of

29. The condition $\beta > \alpha$ will hold if $\beta = 0.1$ (a capacity decline rate of 0.1, or a capacity reserves-to-production ratio of 10) and if $r \leq 0.1$ (a real discount rate of less than 10), provided the rate of price growth α is assumed to be less than r. If $\beta < \alpha$, it can be shown that output always equals capacity (with large initial reserves) and the solution paths are given by equation (36) for all t.

30. The assumption $\beta > \alpha$ implies $\overline{R}/\bar{q} = 1/\beta < R^*/q^* = 1/\alpha$.

31. As a practical matter, the model itself oversimplifies the supply process by ignoring development of new reserves, departures of the price path from a constant rate of growth, and other influences on aggregate supply. Nevertheless, this level of abstraction is useful to derive optimal extraction paths and to illustrate how the solution, which is unambiguous for the firm, becomes ambiguous for the industry.

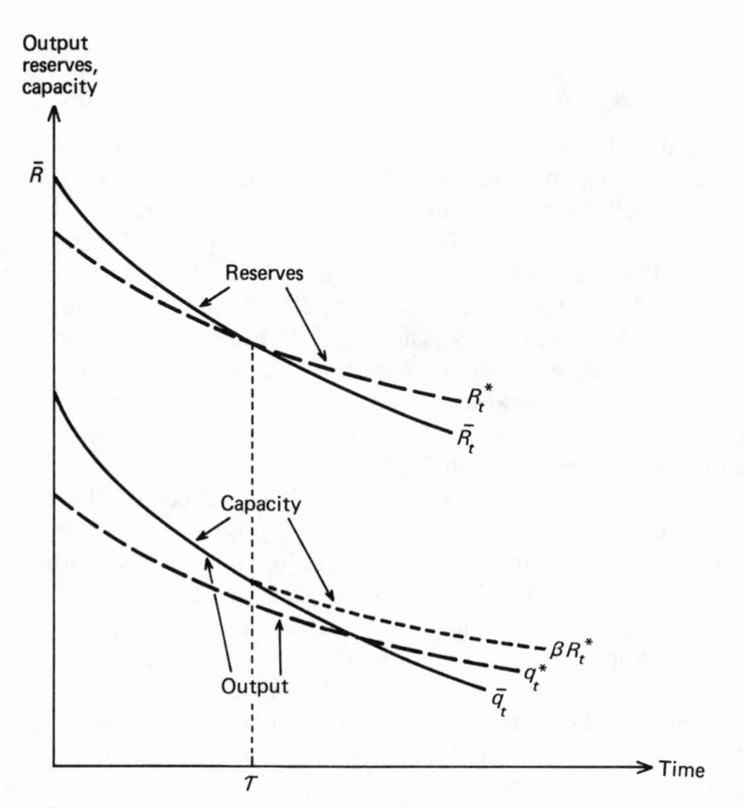

Figure 2-5. Capacity, output, and reserves with constant rate of capacity decline.

capacity constraints. Other dimensions of the problem are discussed in chapters 5 and 6.

Extensions of the Development Model

Little has been said that is specific to the exploration decision and to the product of exploration, new discoveries. Yet exploration is the focus of Peterson's (1978) paper, upon which our discussion of reserve additions is based. Exploration also can be introduced by replacing decisions about reserve additions with decisions about "exploratory effort"—for example, the number of exploratory wells drilled or exploratory drilling footage adjusted for changes in depth. Pindyck (1978c) takes this approach in a model in which exploratory effort is related to gross increases in reserves by a production function that also depends on cumulative past discoveries. Cumulative past discoveries measure the effect of depletion at the discovery stage.

The difference in emphasis between Pindyck's and Peterson's models highlights important practical questions about the specification and measurement of variables in a model of development and exploration. For example, should the supply model focus directly on the response of new reserves to prices and costs (as in Peterson's model), or should the connection be indirect through the responses of exploratory effort with a separate function that relates exploratory effort to new reserves (as in Pindyck's approach)? The latter approach offers the potential for more information, including the influence of technology and input costs, yet it also requires more information. The measure of exploratory effort also is ambiguous, as illustrated by reference to the petroleum industry. Exploratory effort can be measured by seismic activity, number of wells drilled, drilling footage, or by some combination of these; but no single measure is likely to capture the diversity and possibly the intensity of effort. The measurement of depletion effects also is ambiguous. Both Pindyck and Peterson use cumulative discoveries to measure depletion, yet cumulative effort is another possibility. These issues arise again in chapter 6.

At the same time, it is important to separate exploration and development as activities because of dynamic interactions between the decisions involved, including, in particular, the interactions between decisions to develop on the extensive versus the intensive margin. Development on the extensive margin presupposes a new discovery, while development on the intensive margin involves reworking existing deposits. Changes in the cost of reserve additions at the intensive margin relative to the extensive margin—because of depletion of potential resources or changes in input costs and technology—will lead to a reallocation of effort between the two margins. Sorting out these interactions is another issue that must be confronted in empirical analysis of resource supply.

The choice of development and extraction technique is another dimension that influences the stock of developed reserves and extraction capacity, particularly for petroleum. For example, in analyzing development and extraction decisions on the intensive margin, reserves may be extracted by improved recovery techniques or by installation of additional primary recovery capability. The first option effectively augments the stock of producible reserves, whereas the second option does not. Uhler (1979) describes a petroleum supply model in which extraction cost and capacity depend on investment in improved recovery techniques as well as on the size of the reserve. Uhler's analysis can be combined with a description of new discoveries and additions to reserves to obtain a more complete characterization of development and exploration behavior.

BEHAVIORAL IMPLICATIONS AND AMBIGUITIES

Because of its complex, dynamic nature, the theory of nonrenewable resource supply offers few unambiguous predictions about the qualitative effects of changes in economic influences on the behavior of firms. Conclusions depend on specific assumptions about functional relationships and the intertemporal behavior of variables represented in the functions.

Some of the ambiguities may be observed on an intuitive level by noting the offsetting influences represented in the extraction decision. An increase in the current price, for example, may be expected to shift extraction toward the present, and an increase in future prices will shift extraction into the future. Thus, an increase in both current and future prices has an ambiguous effect on the extraction profile unless the change in the price path is more fully specified. Incorporating exploration and development decisions along with extraction decisions further complicates the issue because of offsetting interactions between the decisions. For example, with the possibility of new additions to reserves, the effect of an increase in the current price on the extraction and reserve addition profile becomes ambiguous. On the one hand, the relative decline in future prices reduces the shadow price of new reserves and induces a slowdown in reserve additions; on the other hand, earlier depletion of existing reserves raises the value of new reserves and stimulates exploratory activity. The price change may be expected to increase current output, but the effect on future output depends on whether reserve additions also rise to offset more rapid depletion. Without more specific assumptions, the net effect of the price change cannot be determined.

Behavioral implications of the theory can be investigated in two dimensions. The approach used here is to examine decision paths over time for different specifications of cost relationships and for a particular magnitude of initial reserves and a particular price path. The alternative approach is to consider how decision paths are altered by shifts in the price path or by changes in the initial reserve stock. The first approach is simpler because it involves a comparison of the properties of a particular solution to the producer's optimization problem. The second involves a comparison of the solutions to the optimization problem for different sets of exogenous influences. The latter comparisons are generally inconclusive unless solutions are derived (or simulated) using specific functional forms and price paths.[32] As seen below, even the properties of specific solutions depend critically upon the assumptions made about prices, costs, and reserves.

32. See Toman (1983a) for a discussion and diagrammatic illustration of comparative dynamics properties using a two-period model.

Dynamics of Decision Paths

The dynamics of extraction decisions can be investigated by differentiating the marginal profit rules given by equations (3) and (26) to obtain the following expressions:

$$E_{qq}\Delta q = -r(p - E_q) + \Delta p - E_{qR}(y - q) - E_R \quad (37)$$

$$C_{yy}\Delta y = rC_y - C_{yz}y + C_z + E_R \quad (38)$$

where Δq, Δy, and Δp represent rates of change in variables per unit time. In carrying out the derivation, it is assumed that marginal extraction and development costs are upward sloping, so that $E_{qq} > 0$ and $C_{yy} > 0$.

Each of the terms in the right-hand sides of equations (37) and (38) has an economic interpretation. In equation (37), $-r(p - E_q) < 0$ represents the negative effect that discounting exerts on extraction. Because future revenues are discounted, there is an incentive to extract more rapidly in early periods while reducing output over time. On the other hand, price increases over time will tend to offset this effect, as indicated by term Δp, while a declining price tends to reinforce the discount effect. The term $-E_{qR}(y - q) = -E_{qR}y + E_{qR}q$ represents the marginal effect of depletion over time. Assume that $E_{qR} < 0$, so that depletion raises marginal (along with total) extraction cost. Then $E_{qR}q < 0$ represents the tendency toward declining output because of rising marginal cost, whereas $-E_{qR}y > 0$ represents the offsetting effect on cost and extraction of new reserves. The term $-E_R > 0$ in equation (37) represents a pure time effect: as the time of abandonment draws nearer, the user cost term $\sum^{T} d^{s-t}(-E_R)$ in equation (3) grows smaller and its negative effect on extraction diminishes.

Analogous interpretations apply to the terms in equation (38). Specifically, $rC_y > 0$ represents the positive effect of discounting the cost of future reserve additions, while $-C_{yz}y$ is the marginal depletion effect. If $C_{yz} > 0$, so that marginal cost rises with cumulative additions, then this term captures the negative effect of rising marginal cost on new reserve additions. Both $C_z > 0$ and $E_R < 0$ represent pure time effects as the time of abandonment draws nearer. The former term relates to the decline in development user cost, $\sum^{T} d^{s-t}C_z$, and the latter term relates to the decline in the shadow price of new reserves, $\sum^{T} d^{s-t}(-E_R)$.

In principle, equations (37) and (38) can be solved jointly to determine the time paths of extraction and reserve additions, but in practice these solutions can be derived only by making very strong assumptions about the cost functions. Even ascertaining the signs of Δq and Δy from these equations is complicated by the offsetting influences discussed above. The sign of Δq, for example, will depend on the assumptions made about the various cost function derivatives in equation (37) and the assumptions about the sign and magnitude of the price change Δp.

The shapes of decision paths also depend on interactions between decisions. For example, the direction and magnitude of price changes influence the extraction path, which influences in turn the shadow price of new reserves and the path of reserve additions. Reserve additions also exert a positive influence on extraction, as indicated by the term $-E_{qR}y > 0$ in equation (37). Thus, conclusions about declining output over time implied by a model that considers only the extraction decision will be invalid if reserve additions are large enough to offset the negative effects of depletion and discounting on extraction.

Sensitivity of Conclusions to Price and Cost Specifications

As noted above, conclusions about behavior along an optimal path depend on assumptions made about the price path and the specification of extraction cost. This sensitivity can be illustrated by focusing on extraction behavior with no additions to reserves, so that $y=0$ in equation (37), and by considering the implications of two alternative cost specifications.

The first specification has the general form

$$E(q, R) = E_0(R)E_1(q) + E_2(q) \tag{39}$$

Equation (39) states that extraction cost is multiplicatively separable in the size of reserves and extraction rate. Assume that $E_1(0) = E_2(0) = 0$, so that quasi-fixed cost $E(0, R)$ is zero for all R. Also assume that $E_0' < 0$ (where the prime denotes derivative), indicating that depletion raises the intercept but not the slope of the marginal cost schedule $E_q = E_0E_1' + E_2'$. The second specification is

$$E(q, R) = E_0(R) + E_1(q) \tag{40}$$

which is additively separable in the size of reserves and the extraction rate. Quasi-fixed cost is assumed to be positive and to increase as the stock of reserves declines (so that $E_0 > 0$ and $E_0' < 0$), but depletion does not affect marginal cost, $E_q = E_1'$.

If extraction cost satisfies equation (39), it can be shown that extraction *declines* over time unless the rate of growth in prices is substantial. Intuitively, this follows from the fact that upward shifts in marginal cost due to depletion diminish the marginal profit of extraction over time. In response, and to satisfy the zero-terminal-profit condition (7), output must decline unless price is growing rapidly enough over time to offset the effect of depletion.

If extraction satisfies equation (40), however, the rate of extraction *increases as the terminal time is approached* unless prices are substantially declining. The reason for this surprising result can be deduced from equation (7) and from the fact that depletion raises quasi-fixed and average total extraction cost but not marginal cost. Consequently, as the time of abandonment approaches, extraction can be accelerated to recoup the high level of quasi-fixed cost with no penalty of rising marginal cost. The acceleration of extraction is disadvantageous only if the price declines sufficiently over time. Formal proofs of these assertions are provided in the appendix to this chapter.

Different specifications of the cost components in equation (40) will generate different predictions for extraction. For example, Levhari and Liviatan (1977, p. 187) show that if $E_0(R) = \alpha(\overline{R} - R)$, and if the resource price does not change over time, then extraction will increase over all time periods, not just near the time of abandonment.[33] In contrast, the optimal extraction path will be U-shaped—first declining, then rising—if price is constant and if $E_0(R) = \alpha(\overline{R} - R)^2$, and $E_1(q) = \beta q^2$, so that marginal extraction cost is $E_1' = 2\beta q$ (that is, linear in q) and the depletion effect is $E_0' = -2\alpha R$ (that is, linear in R).

These results indicate the general difficulty of deriving unambiguous predictions from the theory, even at the simplest level of the analysis. The predicted extraction path may rise or fall, or do both alternately, depending on the specification of the extraction cost function and the price path.[34] As a practical matter, however, the additive cost function seems less valid than the multiplicative specification, because the former predicts rising extraction rates while the latter reflects the more realistic dampening effect of rising marginal extraction costs.

CONCLUDING REMARKS

The theory of nonrenewable resource supply behavior outlined in this chapter provides a basis for analyzing the dynamics of supply decisions and the interactions among various stages in the overall supply process. However, the nature of decision paths predicted by the theory is not

33. Levhari and Liviatan's model specifies depletion in terms of cumulative past extraction $x = \overline{R} - R$, where R is the remaining resource stock.

34. Analyses of reserve additions using equation (37) are even more inconclusive: even if separability assumptions analogous to equations (38) and (39) are imposed on the development cost function $C(y, z)$, the sign of Δy remains ambiguous.

easily determined. How decisions respond to economic incentives and constraints depends on specific price and cost trends and on specific properties of cost relationships and capacity constraints. The dynamics of resource supply thus are complex and cannot be assessed from a traditional static perspective or by oversimplified exponential growth rules relating resource prices and output. In particular, long-run price elasticity estimates computed by comparing price and output changes over periods of changing reserves, exploration and development, and technologies have little meaning. Moreover, while some theoretical support can be found for the decline curve approach to modeling supply, this approach is consistent with the theory only under special conditions.

The scope of the basic theory is limited by the underlying assumptions listed at the beginning of this chapter. The next two chapters describe extensions of the theory to relax these assumptions. In applications of the theory, questions also arise concerning (1) the specification of supply relationships and the use of data, (2) differences between the behavior of individual firms and the aggregate of all firms in the industry, and (3) interactions between market supply and demand conditions, as well as institutional constraints. These gaps between theory and application are discussed in chapters 5 and 6.

APPENDIX

This appendix provides formal proofs of assertions made in the text about the relationship among output paths, price paths, and the specification of the extraction cost function.

Suppose first that the extraction cost function satisfies equation (39), with $E_0' < 0$, $E_1' > 0$, $E_2' > 0$, and $E_i'' > 0$ for $i = 0, 1, 2$. Suppose also that $E_1(0) = E_2(0) = 0$. Then $E_q = E_0 E_1' + E_2' > 0$, $E_{qq} = E_0 E_1'' + E_2'' > 0$, $E_R = E_0' E_1 < 0$, and $E_{qR} = E_0' E_1' < 0$; marginal cost is strictly increasing in the rate of output and rises (for any output rate) as reserves decline. In addition, $E(0, R) = 0$ for all R, so that quasi-fixed extraction cost is zero. Note that $E_1(0) = 0$, $E_1' > 0$, and $E_1'' > 0$ imply that $E_1(q)$ is a convex function that passes through the origin. Consequently, $E_1'(q)q - E_1(q) > 0$ for all $q > 0$.

Substituting these derivative formulas into equation (37) and setting $y = 0$ yields

$$\begin{aligned}(E_0 E_1'' + E_2'')\Delta q &= -r(p - E_0 E_1' - E_2') - E_0' E_1 + E_0' E_1' q + \Delta p \\ &= -r(p - E_0 E_1' - E_2') + E_0'(E_1' q - E_1) + \Delta p \quad (41)\end{aligned}$$

Since $p = E_q + (\text{user cost}) > E_q$ along an optimal path by equation (3), the first term on the right side of equation (41) is negative. The second term also is negative since $E_0' < 0$ and $E_1'q - E_1 > 0$, as noted above. Since $E_0 E_1'' + E_2'' > 0$, it follows that $\Delta q < 0$ if $\Delta p \leq 0$. Even if $\Delta p > 0$, output still declines over time provided the increase in price is not too large. In particular, if $\Delta p < r(p - E_0 E_1' - E_2')$, it follows that $\Delta q < 0$. If $\Delta p = rp$, so that price is growing at the rate of discount, then equation (41) implies that

$$\left(E_0 E_1'' + E_2''\right)\Delta q = r\left(E_0 E_1' + E_2'\right) + E_0'\left(E_1'q - E_1\right) \qquad (42)$$

Depending on the relative magnitudes of the derivatives in equation (42), it is possible for output to decline even if $\Delta p/p = r$.

Now suppose that extraction cost satisfies equation (40), with $E_0(R) > 0$, $E_0' < 0$, $E_1' > 0$, and $E_1'' > 0$. Then $E(0, R) = E_0(R) > 0$, so quasi-fixed cost is positive; marginal cost $E_q = E_1'$ is independent of reserves, so that $E_{qR} = 0$; and $E_R = E_0' < 0$. Substituting into equation (37) with $y = 0$ yields

$$E_1''\Delta q = -r\left(p - E_1'\right) + \Delta p - E_0' \qquad (43)$$

It follows from equation (43) that $\Delta q > 0$ if $\Delta p \geq r(p - E_1')$, since $E_0' < 0$. Thus output increases over time, even if $\Delta p < r(p - E_1')$, provided the depletion effect $|E_0'|$ is sufficiently large.

Moreover, it can be shown that $\Delta q > 0$ as the terminal time T is approached, provided price is constant or rising. This follows first from the proposition proved in Kemp and Long (1980a) that the terminal time is finite if there is a quasi-fixed cost of extraction. Consequently, equation (5) implies that $p_T - E_1'(q_T) = p_T - E_q(q_T, R_T) = 0$. Combining this observation with equation (42), it follows that $\Delta q_T > 0$ and, by a continuity argument, that $\Delta q_t > 0$ for t near T.

Thus, with the additively separable specification (40), extraction ultimately increases if prices are rising, and this tendency is strengthened by a larger depletion effect $|E_0'|$. These counterintuitive properties cast doubt on the applicability of this specification. As noted in the text, output is monotonically increasing if $\Delta p = 0$ (no price change over time) and $E_0(R) = \alpha(\overline{R} - R)$. This tendency clearly is strengthened if $\Delta p > 0$. On the other hand, if $E_0(R) = \alpha(\overline{R} - R)^2$, $E_1(q) = \beta q^2$, and $\Delta p = 0$, then it can be shown that the optimal output path is U-shaped. The proof of this assertion involves some rather lengthy computations, however, so we omit it.

3
Joint Products, Externalities, and Technical Change

The basic theory of nonrenewable resource supply presented in chapter 2 has limited practical relevance because of, among other things, the assumptions on which it is constructed. The implications of relaxing three of these assumptions are the topic of this chapter. These are the assumptions that firms produce a single homogeneous output, that there are no interactions between one firm's decisions and other firms' costs, and that resource extraction and development technologies are fixed. Our objective is to describe how the theory of resource supply can be extended when these assumptions are relaxed and to highlight some of the theoretical questions that remain. Relaxing the first assumption recognizes that nonrenewable resources, such as oil and natural gas or minerals from "composite ore," may be discovered and extracted as joint products. The principal issue introduced by relaxing the second assumption is "common access" externalities, which arise (especially with fungible resources like petroleum) when property rights to the resource overlap and recovery is subject to the "rule of capture." The third assumption bears on the intertemporal stability of a firm's costs of finding and extracting resources and on the general question of how to model technical change.

JOINT PRODUCTS

Oil and natural gas are often found together in the same reservoir and can be developed and extracted in varying proportions. Exploration for new petroleum discoveries also can be directed toward oil or gas by intensifying search in areas perceived to be richer in either oil or gas.[1] Similarly, the proportion of individual minerals in a unit of extracted composite ore may vary from one deposit to another, and different proportions of mineral products may be obtained from the ore by changing the way in which it is processed. Where two or more resources are found together, decisions regarding the supply of one resource are linked to decisions regarding the others.

To investigate the implications of joint products for supply behavior, the model developed in chapter 2 is extended to two jointly extracted resources.[2] Consider two resources q_1 and q_2 with a cost of extraction given by a function $E(q_1, q_2, R_1, R_2)$ that depends on the rates of output (q_1, q_2) and the reserve levels (R_1, R_2) of the two resources. It is assumed that depletion of either resource increases the cost of extracting any particular combination of the two resources.

Let p_{1t} and p_{2t} denote the prices of the two extracted resources at date t. Applying the argument developed in chapter 2, the marginal profit rule for positive extraction of the joint products is

$$p_{it} = E_{qi}(q_{1t}, q_{2t}, R_{1t}, R_{2t}) + \sum_{s=t+1}^{T} d^{s-t}[-E_{Ri}(q_{1s}, q_{2s}, R_{1s}, R_{2s})] \tag{1}$$

for $i = 1,2$ and $t = 0,1,\ldots,T-1$, where E_{qi} and E_{Ri} represent derivatives with respect to q_i and R_i. At the terminal time T, the marginal profit rule reduces to

$$p_{iT} = E_{qi}(q_{1T}, q_{2T}, R_{1T}, R_{2T}), \qquad i = 1,2 \tag{2}$$

These relationships are the analogues of the marginal profit rules given by equations (3) and (5) in chapter 2 in the single-product case. A zero terminal profit condition analogous to equation (7) in chapter 2 de-

1. The extent of directionality is an open empirical question, though the principle that firms may alter the probability of finding oil versus gas is less controversial than once believed.

2. This model is a dynamic version of the static joint products model presented in Erickson and Spann (1971). Erickson and Spann's model, which they apply to petroleum discoveries rather than extraction, uses a production function relating discoveries to exploratory effort. To preserve symmetry with the discussion in chapter 2, we recast their model using a dual cost function approach, with cost depending on reserve stocks to capture the dynamic effects of depletion. As indicated below, a static model provides an incomplete description of supply behavior.

termines the optimal terminal time:

$$d^T[p_{1T}q_{1T} + p_{2T}q_{2T} - E(q_{1T}, q_{2T}, R_{1T}, R_{2T})] = 0 \quad (3)$$

Joint products at the exploration and development stages can be brought into the picture in a similar way. Let y_1 and y_2 denote additions to reserves of the two resources, and suppose that the cost of additions is given by a cost function $C(y_1, y_2, z_1, z_2)$ where z_1 and z_2 are cumulative reserve additions of the two resources. Suppose that C has properties analogous to those of the extraction cost function E. In particular, suppose that depletion of the undeveloped resource base for either resource—that is, a rise in either z_1 or z_2—raises the joint cost of reserve additions. Accordingly, along with the extraction profit rules given by equations (1) and (2), we have the marginal profit rules for reserve additions:

$$y_{it} \geq 0, \quad C_{yi}(y_{1t}, y_{2t}, z_{1t}, z_{2t}) + \sum_{s=t+1}^{T} d^{s-t}C_{zi}(y_{1s}, y_{2s}, z_{1s}, z_{2s})$$

$$\geq \sum_{s=t+1}^{T} d^{s-t}[-E_{Ri}(q_{1s}, q_{2s}, R_{1s}, R_{2s})], \quad t = 0, 1, \ldots, T-1 \quad (4)$$

$$y_{iT}C_{yi}(y_{iT}, y_{2T}, z_{1T}, z_{2T}) = 0 \Rightarrow y_{iT} = 0 \quad (5)$$

These expressions relate the opportunity cost of reserve additions (including user cost, $\Sigma d^{s-t}C_{zi}$) to their shadow prices, which in turn are equal to the user costs of extraction, $\Sigma d^{s-t}(-E_{Ri})$. Like their analogues in chapter 2, equations (4) and (5) above are written in "complementary slack" form to allow for the possibility that reserve additions are not made in periods when the opportunity cost exceeds the shadow price.[3] With reserve additions, the zero terminal profit condition (3) must be amended to

$$d^T[p_{1T}q_{1T} + p_{2T}q_{2T} - E(q_{1T}, q_{2T}, R_{1T}, R_{2T}) - C(y_{1T}, y_{2T}, z_{1T}, z_{2T})] = 0 \quad (6)$$

Capacity constraints on extraction rates also can be introduced, as in the derivation of equation (34) in chapter 2, by assuming that extraction capacities $F_1(R_1, R_2)$ and $F_2(R_1, R_2)$ depend on developed reserves and by appropriately modifying the extraction user costs.

3. As in chapter 2, the extraction marginal profit rules can be expressed the same way to allow for temporary cessation of extraction.

Implications of the Joint Products Model

The marginal profit rules and zero terminal profit condition can be solved in principle (using the recursive procedure described in chapter 2) to obtain dynamic supply functions for extracted outputs and reserve additions:

$$\begin{aligned} q_{1t} &= f_{1t}(p_{1t},\ldots,p_{1T};\, p_{2t},\ldots,p_{2T};\, R_{1t}, R_{2t}, z_{1t}, z_{2t}) \\ q_{2t} &= f_{2t}(p_{1t},\ldots,p_{1T};\, p_{2t},\ldots,p_{2T};\, R_{1t}, R_{2t}, z_{1t}, z_{2t}) \end{aligned} \quad (7)$$

$$\begin{aligned} y_{1t} &= g_{1t}(p_{1t},\ldots,p_{1T};\, p_{2t},\ldots,p_{2T};\, R_{1t}, R_{2t}, z_{1t}, z_{2t}) \\ y_{2t} &= g_{2t}(p_{1t},\ldots,p_{1T};\, p_{2t},\ldots,p_{2T};\, R_{1t}, R_{2t}, z_{1t}, z_{2t}) \end{aligned} \quad (8)$$

These equations, and the marginal profit rules themselves, indicate a source of interdependence among decisions that was not encountered in chapter 2: dynamic price and cost interactions among joint products both within and across stages of the supply process. As indicated by equations (1) and (4), the user cost of extraction for each resource depends on extraction decisions and reserve stocks of other joint products. Reserve additions for each resource also depend on cumulative additions of other jointly discovered and developed reserves, through their influence on the shadow price of new reserves.

Consequently, as indicated by equation (7), the rate of extraction of each resource depends on current and future prices of other joint products. This cross-price relationship arises directly through the effect of relative output prices on the mix of extracted outputs, and indirectly through the effect of relative prices on the proportions of reserves added. As indicated by equation (8), the dependence of extraction on the prices of all joint products influences the shadow prices of new reserves, so that additions to reserves also depend on relative output prices. In addition, extraction decisions depend on the size of existing reserve stocks for all jointly produced resources through their influence on extraction user costs. Extraction also depends on cumulative additions to all reserves, because these variables influence the shadow prices of new reserve additions and the shadow prices affect the proportion of each resource in new reserve additions. Exploration and development decisions depend in a similar way on both the size of existing reserve stocks and on cumulative additions to reserves.

Predictions about the cross-price effects on extraction and reserve additions indicated by equations (7) and (8) are ambiguous in the absence of more specific assumptions because of offsetting influences. For example, an increase in the current relative price of natural gas will induce an increase in the rate of extraction of natural gas when depletion effects are ignored. This substitution effect is illustrated in figure 3-1a, which shows isocost curves for fixed reserve stocks and two levels of total outlay. For any level of outlay, an increase in the current relative price of gas will induce a substitution away from oil. For reasonably well behaved cost functions, the change in the relative price will induce higher gas output and lower oil output, even when changes in scale are accounted for. However, the total effect of the gas price change on oil extraction also depends on the degree to which the increase in gas output induces change in future oil production costs, and on the effect of adjustments in oil extraction on future extraction costs. These depletion effects are illustrated in figure 3-1b and alter extraction rates in ways that may be

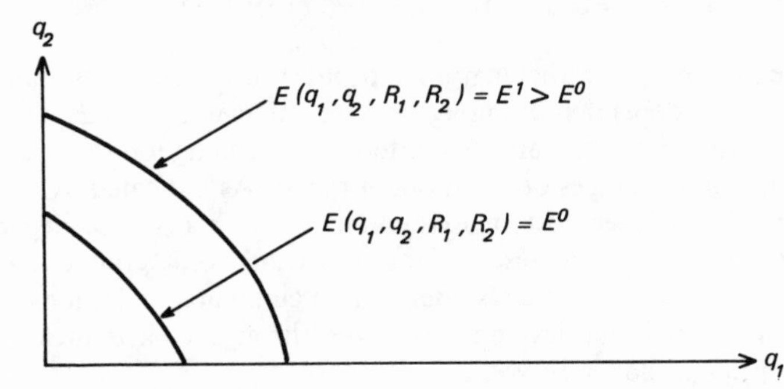

Figure 3-1a. Properties of joint products extraction cost function: Isocost curves for fixed resource stocks, increasing outlay.

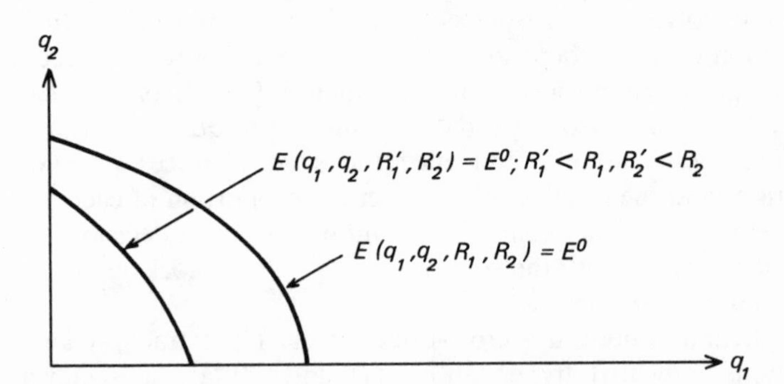

Figure 3-1b. Properties of joint products extraction cost function: Effect of depletion on output possibilities with fixed expenditures.

similar or opposite to the atemporal substitution effects discussed above. Whether the increase in the price of natural gas is temporary or permanent also has a bearing on the outcome for both gas and oil production, since future prices influence relative user costs.

Changes in relative costs over time also can alter the predicted outcome for supply behavior. For example, a change in the relative costs of reserve additions will lead to a change in relative proportions of extracted outputs as well as in reserve additions. Cost changes may offset or reinforce the influence of relative price changes. Relative cost shifts may result from institutional and market factors such as regulatory constraints, technical change, and changes in capital and materials costs. Changes in the relative value of certain reserves over others (for example, because of increased risk of expropriation of foreign oil reserves) will also lead to changes in the proportion and scale of activities throughout the supply process.

Processing and Storage of Joint Products

For many nonrenewable resources, refining and processing of extracted raw materials to separate constituent joint products is another important stage of the supply process. In addition, the possibility (and sometimes, the necessity) of storing some or all of the constituent products requires a distinction between resource extraction that adds to inventories and sales that reduce inventories.

To illustrate the implications of processing and storage for supply behavior, let q denote the quantity of extracted raw "ore," and let q_i, $i=1,\dots,n$ denote the products obtained by processing the ore.[4] Let R denote the level of ore reserves, and assume that the cost of ore extraction is given by $qh(R)$, with $h'(R)<0$. In words, the marginal (and average) extraction cost $h(R)$ for ore is independent of the extraction rate but rises as reserves fall. Let $E_i(q_i, q)$ denote the cost of separating product i from the composite ore at the processing stage. Processing cost is assumed to be independent of production rates for other composite products and to vary inversely with the total amount of ore available: the larger the quantity of raw material, the lower the processing cost for any of the constituent joint products contained in the raw material.

Starting first with the case of zero storage, so that production and sales are identical, let $E_{iq}=\partial E_i/\partial q<0$ denote the rate at which processing cost for the ith product rises as the quantity of extracted ore available is reduced; let $E_i'=\partial E_i/\partial q_i>0$ denote marginal processing cost; and let

4. The discussion that follows is based on Pindyck (1982). We have changed Pindyck's notation somewhat to conform to our own.

p_{it} denote the market price of the ith product at time t. Under these conditions, an optimal program of ore extraction and processing is described by the conditions

$$-\sum_{i=1}^{n} E_{iq}(q_{it}, q_t) = h(R_t) + \sum_{s=t+1}^{T} d^{s-t}[-q_s h'(R_s)], \qquad t = 0,\ldots,T-1 \tag{9}$$

$$-\sum_{i=1}^{n} E_{iq}(q_{iT}, q_T) = h(R_T) \tag{10}$$

$$p_{it} = E_i'(q_{it}, q_t) \qquad i = 1,\ldots,n; \qquad t = 0,\ldots,T \tag{11}$$

In equations (9) and (10) the term $-\sum_{i}^{n} E_{iq}$ measures the decline in total processing cost from an additional unit of extracted ore. This term is the shadow price of new ore to the firm.[5] The term $\Sigma d^{s-t}(-qh')$ in equation (9) is the user cost of ore extraction and measures the present value of future cost increases from running down reserves. The condition stated in equation (11), which equates marginal processing costs (given an optimal rate of ore extraction) and the market prices of the constituent products, reflects the absence of user cost associated with processing ore when processing costs are assumed to be independent of the stock of reserves R and processing costs for each product are independent of the processing rates of other products. The second assumption can be relaxed by modifying equation (11) to include the term $\sum_{j \neq i} \partial E_j / \partial q_i$ on the right-hand side, thus capturing how increases in q_i raise processing costs for other products q_j.

Adding an inventory decision to the model requires an additional marginal profit rule. Equations (9) through (11) continue to describe the extraction and processing decisions, while the relation between inventories and sales is implicitly defined by the arbitrage equation

$$(p_{i,t+1} - p_{it}) = rp_{it} + k_i \tag{12}$$

where r is the discount rate and k_i is marginal storage cost for the ith product (assumed to be positive and constant). Equation (12) states that for inventory accumulation to be optimal, the capital gain from the last increment to storage must equal the sum of interest cost rp_{it} (the forgone earnings from current sale and reinvestment of proceeds) and storage cost k_i.

5. Pindyck (1982) notes that the shadow price can be interpreted as the "transfer price" of ore, or the amount that nonintegrated firms would pay or receive when buying or selling the ore.

Introducing the storage decision highlights the additional influence that the balance between current revenue prospects and inventory capital gains exerts on supply behavior. The model also provides a point of departure for analyzing more complex production processes in which firms may sell their outputs, store them for future sale, or use them as inputs to the production of other resources. In developing oil reservoirs with associated gas, for example, firms may choose to market the gas immediately, hold the gas in storage for sale in periods of peak demand, or reinject the gas into the reservoir to supplement the natural reservoir drive and increase total oil recovery.

Empirical Significance of Joint Products

The joint products model with dynamic cross-price effects introduces an additional element of realism by recognizing that many resources are found together in the earth's crust and are jointly developed and extracted. In general, the volume of discoveries and the rate of extraction of one joint product are expected to depend on the prices of others. The strength of this relationship is, of course, an empirical question. For oil and natural gas, in particular, it is reasonable to characterize exploratory drilling as less than perfectly "directional," in which case cross-price effects are potentially important. The extreme hypothesis of no cross-price effects requires not only perfect drilling directionality, but also sufficient nonassociated oil or gas resource potential for an increase in the price of oil or gas to lead to discovery of only nonassociated reservoirs.[6]

The potential relevance of joint products in oil and gas supply is strengthened when uncertainty is brought into the picture. Because firms do not know the precise location and composition of petroleum deposits but have only an estimate of how deposits are distributed, there is an important distinction between *ex ante* decisions and *ex post* outcomes. Expectations of discovering only oil or only gas will be correct at best on average. Consequently, *ex post* data should reveal some cross-price effects, according to this view, even if drilling can be directed entirely toward oil or gas on average.

COMMON PROPERTY EXTERNALITIES

Common property externalities in the petroleum industry have been an important subject of scholarly interest and public debate for many years,

6. Note also that the other extreme of complete randomness in discovery proportions implies an equivalent response of oil or gas discoveries to changes in either resource's price, with the discovery mix being determined entirely by relative geological abundance of the two resources rather than by any ability to discriminate among potential prospects.

and numerous regulations have been imposed on the industry to limit the adverse effects of cost spillovers. Oil and natural gas resources are fungible in underground deposits and property rights to extract the resources may overlap. When more than one firm is extracting from the same reservoir, depletion effects depend on the sum of individual firms' decisions.[7] Acting alone, each firm will be influenced in its decision making by its expectations of other firms' activities, through the effect of those activities on its costs. Similar external cost spillovers may arise for any nonrenewable resource at the discovery stage when several firms are exploring simultaneously in a region where property rights have not been clearly defined. Depletion of the total resource base—and hence each firm's cost of finding new reserves—will depend on the sum of reserve additions by all firms.

As an illustration of how these externalities affect decisions of individual firms, consider a situation in which there are $N > 1$ firms extracting a common property nonrenewable resource.[8] The cost of extraction to firm i satisfies the relation $E^i(q^i, q, R) = q^i h^i(q, R)$, where q^i is the output of firm i; $q = \sum_{i=1}^{N} q_i$ is the sum of the firms' outputs; R is the total stock of resource available to all firms; and $h^i(q, R)$ is firm i's unit cost of extraction. Given an initial reserve stock of $\overline{R}$, total reserves satisfy the relation

$$R_{t+1} = R_t - q_t = R_t - \sum_{i=1}^{N} q_t^i, \qquad t = 0, 1, \ldots \tag{13}$$

Each firm is assumed to choose an extraction plan that maximizes the present value of its net revenues, given expectations of prices and of other firms' extraction plans.[9] Under the assumption that these expectations are correct, each firm's marginal profit rule is given by

$$p_t = \left[h^i(q_t, R_t) + q_t^i h_q^i(q_t, R_t)\right] + \sum_{s=t+1}^{T} d^{s-t}\left[-q_s^i h_R^i(q_s, R_s)\right] \tag{14}$$

7. See McDonald (1971) for a further discussion of this phenomenon. It is interesting to note that firms' operations need not be physically adjacent for common property problems to arise: resource fungibility may extend over several miles, and apparently unrelated reservoirs in fact may be connected.

8. The discussion that follows is based on Aivazian and Callen (1979). For other references, mostly using the simple Hotelling model, see Salant (1982). The original notation used by Aivazian and Callen has been changed slightly to conform to our own. We have also converted their representation of depletion effects in terms of cumulative extraction to a representation in terms of remaining reserves, and we have allowed for variation in unit extraction costs across firms.

9. Formally, this framework is a dynamic analogue of the static Nash–Cournot noncooperative equilibrium model of oligopoly. See Heal (1982) for a general discussion of this framework in the relation to externalities. The weaknesses of the approach are discussed below.

In equation (14), h_q^i is the rate at which unit extraction cost increases with an increase in the total extraction rate (or with an increase in the extraction rate of a single firm, holding other extraction rates fixed) and h_R^i is the rate at which unit cost rises with depletion of total reserves. The term $h^i + q^i h_q^i = \partial E^i / \partial q^i$ is firm i's marginal extraction cost, and $\Sigma d^{s-t}(-q_i h_R^i) = \Sigma d^{s-t}(-\partial E^i / \partial R)$ is the user cost of extraction to firm i. This user cost measures the present value of future increases in firm i's cost attributed to its own decisions, given the actions of other firms.

Equation (14) has the same general interpretation as the marginal profit rule given by equation (3) in chapter 2 for a single firm, except that both marginal operating cost and marginal user cost now depend on the actions of all N firms. If the system of individual marginal profit rules (and the zero terminal profit conditions) could be solved for optimal decisions, the general form of the individual extraction decision would be

$$q_t^i = f_t^i(p_t, p_{t+1}, \ldots, p_T; R_t; \bar{q}_t^i, \ldots, \bar{q}_T^i) \tag{15}$$

where the $\bar{q}^i = \sum_{j \neq i} q^j$ measure the total outputs of the other firms. These variables measure the effect of other firms' supply decisions through the cost spillovers. Because of the dynamic interactions between extraction and reserve replacement discussed in chapter 2, these extraction cost spillovers will also influence firms' incentives to explore and develop new reserves, even if there are no externalities at these stages.

Similar interactions emerge from common property externalities at the discovery stage. To focus on this stage, suppose that extraction cost $E^i(q^i, R^i)$ is independent of other firms' decisions, but that firm i's cost of reserve additions $C^i(y_i, z)$ depends on cumulative reserve additions by all firms.[10] Reserves and cumulative additions are given by

$$\begin{aligned} R_{t+1}^i &= R_t^i - q_t^i + y_t^i, \qquad t = 0, 1, \ldots; \qquad R_0^i = \bar{R}^i \\ z_{t+1} &= z_t + y_t = z_t + \sum_{i=1}^{N} y_t^i \end{aligned} \tag{16}$$

where $y = \Sigma y^i$ is the sum of the firms' additions to reserves. In this situation, the marginal profit rule for each firm's extraction decision is given by equation (3) in chapter 2 because of the assumption of no extraction externalities. The marginal profit rule for reserve additions is

$$C_{yi}^i(y_t^i, z_t) + \sum_{s=t+1}^{T} d^{s-t} C_z^i(y_s^i, z_s) = \sum_{s=t+1}^{T} d^{s-t}[-E_R^i(q_s^i, R_s^i)] \tag{17}$$

10. Peterson (1978) presents a similar analysis, which assumes that each firm's cost depends separately on its own cumulative additions and on cumulative additions by all firms.

ignoring, for simplicity, the possibility that zero reserve additions may be optimal.

Equation (17) is generally the same as equation (26) in chapter 2, the expression with no cost externalities in reserve additions, except that exploration decisions of each firm now are interrelated through the effect of cumulative past discoveries on each firm's marginal exploration cost and exploration user cost. The exploration supply functions determined by equation (17) can be expressed by the equality

$$y_t^i = g_t^i(p_t, p_{t+1}, \ldots, p_T; R_t^i; z_t^i + \bar{z}_t^i) \tag{18}$$

where $z_t^i = \sum_{s=0}^{t-1} y_s^i$ represents cumulative additions by firm i, and $\bar{z}_t^i = \sum_{j \neq i} z_t^j$ represents cumulative additions by other firms. These spillovers also will influence extraction decisions indirectly through their impact on the user cost of depleting existing reserves.

Implications of the Theory

As noted above, one implication of the theory is that "strategic" interactions among individual firms' decisions that arise from common property externalities at any one stage of the supply process also spill over to other stages, given the dynamic interconnections among the stages. Another important implication is that total resource extraction and reserve additions will tend to be larger under conditions of "free entry"—unrestricted common property exploitation—than would be the case if the affected firms formed a "unitized" management plan to internalize the cost spillovers.[11] The tendency toward overextraction can be seen most clearly by examining equation (14) in the special case in which unit cost $h = h(R)$ depends only on total reserves and is the same for each firm. In this case, each firm's equilibrium extraction path will be identical, with $q_t^i = q_t/N$ for all $t \geq 0$. Summing equation (14) across $i = 1, \ldots, N$ and dividing both sides by N yields[12]

$$p_t = h(q_t, R_t) + (1/N) \sum_{s=t+1}^{T} d^{s-t}[-q_s h'(R_s)] \tag{19}$$

As N grows large in equation (19), the user cost term goes to zero and each firm ignores its contribution to depletion effects. Each firm extracts

11. In petroleum extraction and development, unitization refers to an agreement among firms with property rights to a single reservoir to form a joint management (McDonald, 1971).

12. Aivazian and Callen's (1979, pp. 86–87) formulas are slightly different because they also assume that producing firms are oligopolistic rather than competitive.

at the point where price and unit operating cost are equal.[13] Because user cost is ignored in this extreme free-entry scenario, total output exceeds the amount under efficient unitized (or single-firm) management. This corresponds to the classic "problem of the commons" (Heal, 1982).

Similar conclusions can be drawn about free entry at the exploration stage. Suppose that each firm's exploration cost satisfies the expression $C(y^i, z) = y^i k(z)$, where the unit cost of exploration $k(z)$ is the same for all firms. Substituting this cost specification into equation (17), summing over $i = 1, \ldots, N$, and dividing by N yields

$$k(z_t) + (1/N) \sum_{s=t+1}^{T} d^{s-t} y_s k'(z_s) = (1/N) \sum_{i=1}^{N} \sum_{s=t+1}^{T} d^{s-t}\left[-E_R^i(q_s^i, R_s^i)\right] \qquad (20)$$

As N gets larger, the term expressing the user cost of exploration on the left-hand side of equation (20) vanishes, and total new reserve additions satisfy the condition that unit finding cost equals the average shadow price (averaged across all firms). Once again, user cost is ignored in the aggregate and new reserve additions are larger than the efficient unitized amount.[14]

These observations also suggest that free entry magnifies the effect of price and cost changes relative to unitization. The practical significance of this supposition becomes apparent in empirical analyses of aggregate supply behavior when different producing regions are subject to different degrees of common property externalities and when the degree of external cost spillovers is changing over time. For example, common property extraction externalities have had greater historical significance in parts of Texas than in other producing regions, and their significance in Texas has changed as unrestricted practices were replaced first by state conservation regulations on drilling and extraction and later were supplanted with compulsory unitization.

Weaknesses in the Theory

Although the models discussed above are internally consistent and yield predictions that generally conform to the classic view of the "commons problem," several questions remain about the validity and completeness

13. We are assuming in equation (19) that extraction rates satisfy this internal optimum condition and are not constrained at a capacity level.

14. See also the simulation experiments reported by Peterson (1978, pp. 242–248). In his analysis, Peterson considers an exploration cost specification in which unit cost rises with the rate of additions as well as with cumulative additions.

of the theory. To begin with, the models predict that total extraction and reserve additions by all firms rise as the number of firms increases, but they also predict that output and reserve additions by each firm *decline*.[15] Intuitively, this response is due to the fact that resource depletion raises each firm's extraction and exploration user costs, so a firm will react to expectations about other firms' activities by cutting back its own extraction and exploration. However, this theoretical finding is at odds with the widely held view that, faced with higher costs and loss of ultimately recoverable resources through the "rule of capture," petroleum firms respond by increasing their planned extraction and exploration.[16] According to the model, a firm will respond to external cost spillovers by increasing extraction and exploration only if the spillovers are so large that the firm is moved to discount the future entirely, ignore user cost, and rapidly abandon its activities.

At a deeper level are questions about how expectations are formed and how firms respond to rivals' activities. As the theory has been developed, the presumption is maintained that each firm chooses an optimal sequence of activities, given correct expectations about the *sequences* of rivals' decisions. The theory predicts equilibrium decisions in the sense that no firm has an incentive to depart from its previously determined plan. In particular, the equilibrium hypothesis implies that no firm revises either its expectations of others' actions or its own decisions over time as it observes the decision paths of its rivals.[17]

In practice, of course, a firm cannot have perfect foresight of its rivals' actions. Instead, the firm must base its forecast of rivals' actions on currently available information and past behavior, and update these forecasts over time as circumstances require. Changes in expectations and the nature of reactions thus play a critical role in determining a firm's responses. However, the theory described in this section offers no hypotheses about how expectations are formed.[18]

A related issue is the question of the content of these expectations. The theory described above presumes that expectations are formed about the *paths* of rivals' decisions and that, given these expectations, each firm chooses its own path of decisions (at $t = 0$) to which it adheres over time. This approach overlooks the possible incentive for firms to deviate from the hypothesized equilibrium paths in subsequent periods, when reserve stocks and costs differ from conditions in the initial period. In other

15. See Toman (1983a) for a formal treatment of this issue.

16. Davidson (1963); McDonald (1971, chapters 3 and 5).

17. In the theory of noncooperative games, this equilibrium concept is known as a dynamic Nash equilibrium in simple strategies; it is a direct dynamic analogue of the Cournot theory of duopoly. See Friedman (1977) and Toman (1982a) for discussion of these points.

18. Friedman (1977, chapter 5; 1983) discusses game models with explicit hypotheses about expectations and reactions.

words, the approach does not require that decision strategies and expectations be intertemporally consistent. Instead, the models describe the actions of firms that are obligated by some binding precommitments to follow particular sequences of decisions. Though long-term contracts in industries like natural gas might approximate such binding precommitments, fixed long-term agreements are not pervasive in resource extraction, and those that exist typically contain some degree of flexibility. In the absence of binding commitments, rational producers cannot be presumed to place their faith in expectations of fixed strategies by rival firms as required by the theory, and the hypothesized equilibrium based on these expectations may break down.

An alternative approach that avoids the consistency problem is to posit that (1) firms have expectations about their rivals' decision *rules*, rather than decision *paths*; and (2) the decision rules are in "feedback" form, so that they specify optimal decisions for any reserve stock sizes, current price, and costs. Feedback decision rules will be credible because firms will have no incentive to depart from them (by definition). Consequently, the "closed-loop" equilibrium based on expectations of these decision rules will be dynamically consistent, while an "open-loop" equilibrium based on expectations of fixed strategies may not be.[19] Eswaran and Lewis (1982) show that in some cases the two equilibrium concepts give quite different results, while in other cases the results are closely similar. Thus the general question of how significant the dynamic consistency problem is in common property resource models remains open. The question is important in part because closed-loop equilibria typically are far more difficult to analyze. Therefore, to the extent that the two equilibrium concepts yield similar conclusions, the open-loop approach has the advantage of greater tractability.

Finally, the theory described in this section also presumes that each firm reacts to expectations of rivals' decisions but does not attempt to influence rivals' behavior through its own actions. Thus, for example, there is no scope in the model for rapid depletion by one firm as a "threat" to induce rival firms to abandon their operations or to deter entry by new producers. Recent developments in game theory have addressed these issues,[20] but applications of these developments to common property resource supply have been limited. An exception is Toman

19. Incorporating credible strategies leads to the concept of a "perfect" Nash equilibrium. See Reinganum and Stokey (1981) for a discussion of this concept and an application to a simple common property resource model without depletion effects. The issue of precommitment also arises in connection with a related "dynamic inconsistency problem" discussed in chapter 5. "Feedback" decisions rules also figure prominently in the analysis of supply behavior under uncertainty, as indicated in chapter 4.

20. See, for example, Friedman (1979a, 1979b).

(1983c), who uses a simple model to illustrate that when entry deterrence is brought into the analysis, predictions of overexploitation at both the firm and the aggregate level are obtained. These predictions are consistent with the stylized facts of common property petroleum externalities, in contrast to the predictions of models that ignore strategic entry and exit decisions.

Information Externalities

The discussion of common property externalities has focused on "congestion spillovers" resulting from unrestricted entry. Peterson (1978, pp. 238–239) considers a second type of exploration externality that arises because information is a joint product of exploratory effort. Even when exploratory effort is unsuccessful in finding resources, valuable information is revealed about discovery prospects at and adjacent to the sites explored. Moreover, this information is nonappropriable by the firm investing in the exploration project; its actions inevitably reveal to other firms at least some of the information gained.

This type of externality can be incorporated into a common property exploration model by assuming that discovery costs in a particular region fall as total exploratory effort or discoveries rise, at least to the point where congestion externalities begin to outweigh information spillovers. A negative relationship between a firm's exploration costs and the exploration activities of other firms suggests that, under free entry, both individual and aggregate exploration will be larger than under efficiently unitized conditions.[21] Whether this is in fact the case depends on how firms respond to the externality: if they choose to wait for other firms to assume the cost burden of acquiring the common pool of exploration, exploration may be reduced (Stiglitz, 1975). In any event, information spillovers may impart a bias to the location of search activity; specifically, a preference for exploration on the intensive margin where there is less uncertainty about discovery prospects.

Information externalities also arise at the preliminary stage of assessing potential resource-bearing sites and contracting to acquire mineral rights. Depending on the nature of the contracting process—open auction, sealed bid, bilateral negotiation—and the motives of participants (speculation on leasing rights versus demand for new reserves as an input to extraction), either too much or too little may be invested in information gathering.[22] In practice, the significance of information spillovers in

21. This tendency, if it is correct, would offset part of the weakness in the theory discussed in the previous subsection.

22. See Peterson and Fisher (1977, pp. 697–699) for further discussion and references; see also Gilley and Karels (1981) for an empirical study of bidding systems.

the United States has changed as information about the resource base has grown and activity has shifted toward federal lands, particularly offshore prospects, with various systems for allocating mineral rights. These changes pose an additional complication for empirical analysis of resource exploration and for prediction of future trends based on past experience.

TECHNICAL PROGRESS

Technical change that is wholly exogenous to firms' decisions is frequently incorporated into conceptual models by a simple time-related shift parameter in the cost functions, as indicated in equation (14) in chapter 2. This practice raises important practical questions concerning the specification of empirical models and the measurement of technical change, but it raises no new theoretical issues. Of interest in this section is endogenous technical progress, where depletion and rising costs motivate firms to invest in improved technologies for extraction and reserve additions.

One approach to modeling endogenous technical change is to apply the methodology of Kamien and Schwartz (1969) to a resource mining firm. Consider first the extraction decision alone, and suppose that extraction cost is given by $E(q, R, k)$ where $k \geq 0$ represents cumulative investment in "knowledge" or in "improving the state of the art." Defining and measuring this variable poses serious difficulties in practice, but from a conceptual standpoint a more precise definition of k is not required. Suppose that $E_k = \partial E/\partial k < 0$, so that increases in knowledge lead to a reduction in extraction cost. Assume further that $E_{Rk} > 0$, so that increases in knowledge reduce the magnitude of the depletion effect $E_R < 0$. Let $u > 0$ be the rate of investment in knowledge, so that k and u satisfy

$$k_{t+1} = k_t + u_t, \qquad t = 0, 1, \ldots; \qquad k_0 = 0 \tag{21}$$

Let $D(u)$ denote the cost of investment in knowledge.

A present-value-maximizing extraction firm will choose paths of resource extraction and investment in knowledge so that the price of the extracted resource is equal to the marginal opportunity cost of extraction, and the marginal cost of newly acquired knowledge is equal to its shadow value (ignoring corner conditions). The marginal profit rule given by equation (3) in chapter 2 for extraction remains valid, except that extraction user cost $\Sigma d^{s-t}[-E_R(q_s, R_s, k_s)]$ is reduced by investment in better extraction techniques, given $E_{Rk} > 0$. The marginal profit rule for this investment is given by

$$D'(u_t) = \sum_{s=t+1}^{T} d^{s-t}[-E_k(q_s, R_s, k_s)] \tag{22}$$

where $D'(u)$ is the marginal cost of investment and the right-hand side is the shadow value of investment. This shadow value is equal to the present value of future decreases in extraction cost as a result of investment in improved technology.

This formula is similar to the marginal profit rule given by equation (27) in chapter 2 for reserve additions; indeed, improved technology substitutes (imperfectly) for new reserves. It follows that the discussion in chapter 2 of dynamic interactions between extraction and reserve additions also applies here to interactions between extraction and investment in knowledge. These investments ultimately depend on both current and expected future prices of the extracted resource, and (negatively) on the existing reserve stock. Moreover, extraction decisions depend (positively) on the stock of accumulated knowledge.

This approach can be applied in a similar way to technical progress in discovering and developing reserves, with a cost of reserve additions function $C(y, z, k)$ that depends on accumulated knowledge. In this case equation (27) in chapter 2 applies to reserve additions, but the user cost term $\Sigma d^{s-t}C_z(y, z, k)$ will be negatively related to k if $C_{zk} < 0$. The marginal profit rule for knowledge investment becomes

$$D'(u_t) = \sum_{s=t+1}^{T} d^{s-t}[-C_k(y_s, z_s, k_s)] \tag{23}$$

where $-C_k(y, z, k) > 0$ measures the decline in exploration and development cost from increased knowledge, so that the right-hand side of equation (23) is the shadow value of the investment. In this situation both extraction decisions—given by equation (3) in chapter 2—and reserve additions will be intertemporally related to investment in improved techniques.

Discretionary investment is not the only source of endogenous technical change or endogenous shifts in cost. The acts of extracting, finding, and developing new reserves convey information about the size of deposits, about their geological and geophysical characteristics, and about ways to modify existing techniques or to use them more efficiently.[23] Like the information externalities discussed above, the effect of information acquired from past activities can be modeled (as a rough first approximation) by assuming that costs are negatively related to cumulative production, reserve additions, or exploration and development effort. How "learning by doing" influences costs and behavior raises additional questions about the interpretation of depletion effects and the ability to separate these effects from technical change in empirical studies of supply behavior.

23. See Arrow (1962) for an early yet timely treatment of this point.

CONCLUDING REMARKS

The topics discussed in this chapter further emphasize the inherently dynamic nature of nonrenewable resource supply with interactions among joint product decisions, among the strategies of individual firms subject to common property externalities, and between resource depletion and technical progress. The discussion highlights important practical problems concerning the measurement of tradeoffs between joint products decisions and the influence of technical change. We have also indicated gaps within the conceptual analysis of common property supply decisions which hinder understanding of this issue.

Each of the topics has a counterpart on the demand side. Nonrenewable resources, particularly oil and natural gas, are joint products in demand as well as supply. In addition, strategic interactions among economic actors also emerge in markets that depart from the ideal of perfect competition. Finally, technical progress includes development of substitutes for depletable resources as well as more efficient ways of depleting them. These market-level issues are discussed further in chapter 5. Before taking up these issues, we first consider another extension of the basic theory: the introduction of uncertainty. This extension is vital to understanding supply behavior at both the individual and market levels.

4
Uncertainty, Expectations, and Supply Behavior

To extend the theory of supply behavior sketched in chapter 2 toward a more realistic description of actual behavior, the theory must be modified to incorporate an explanation of how uncertainty about prices and costs influences firms' decisions. Future prices are inherently uncertain because of unpredictable shifts in resource demands (for example, shifts due to fluctuations in consumer incomes and tastes or the prices of substitutes), because of changes in market structure (such as the rise in market power of the Organization of Petroleum Exporting Countries), and because of regulatory constraints (such as taxes and direct controls). Costs are inherently uncertain because of imperfect knowledge about the underlying geological and geophysical characteristics of resource deposits, about the size of undeveloped reserve stocks (which influence costs through depletion effects), about technical change, and about the amount of exploratory and development effort required to obtain an increment to reserves.

Much of the literature on resource supply under uncertainty is concerned with the simple model of optimal extraction from a fixed stock of reserves with no depletion effect on costs (that is, costs are independent of remaining reserves).[1] Attention has focused on the implications of an

1. Refer to the section on the "*r* percent rule" in chapter 2.

uncertain total resource stock and on uncertainty about a single discrete shift in future resource demand.[2] The conclusions have limited relevance for analyzing observed market behavior, however, because of the narrow context in which they are derived: depletion effects, reserve additions, and incomplete exhaustion all are ignored. In addition, uncertainty is introduced as a single discrete event, such as a change in the size of the fixed resource base or a once-and-for-all shift in resource demand. In practice, uncertainty is a continuous phenomenon, with expectations about prices and costs in a persistent state of flux.

This chapter is concerned with representing supply behavior under uncertainty by means of models that retain the concepts of depletion effects, incomplete exhaustion, and interactions among stages in the supply process. Particular attention is focused on the application of linear "rational expectations" models to nonrenewable resource supply, and on some of the important gaps in existing theoretical analyses.[3]

MODELING EXPECTATIONS AND DECISIONS UNDER UNCERTAINTY

Deterministic theories of supply behavior presume that firms seek to maximize the present value of net revenue having full knowledge about future prices and costs. When uncertainty is brought into the picture, three additional issues must be addressed: individuals' attitudes toward risk, the way in which their expectations about future events are determined, and their ultimate response to uncertainty. Individual decision makers may be risk averse, risk seeking, or risk neutral.[4] The distinction among the three attitudes toward risk is important because of the corresponding differences in the presumed objectives, and thus the behavior, of the producer. For the most part, the discussion in this chapter assumes that producers are risk neutral, in which case they are presumed to maximize the mathematical expectation of the present value of net revenues.[5] As we will show later in this chapter, risk aversion may

2. For references, see Pindyck (1980, pp. 1204–1206); Dasgupta and Heal (1979, chapter 13); and Peterson and Fisher (1977, pp. 696–700).

3. The subject of intertemporal decision making under uncertainty is inescapably more complex than the material in previous chapters and requires a technical background that may be unfamiliar to nonspecialists. We have attempted to avoid purely technical digressions and have placed some of the more complicated formal arguments in an appendix. Nevertheless, some readers may prefer to focus on the first and last sections.

4. See Arrow (1971) for an introduction to risk preferences.

5. Note the contrast between the expectation of the present value of net earnings and the present value of expected future net revenues.

substantially alter optimal decision rules, but theoretical understanding of risk averse behavior in a dynamic context is far from complete.

An expectations hypothesis is an assumption concerning the way in which expectations about future events are determined. To be meaningful, an expectations hypothesis must postulate that predictions depend on information that is available to the producer, while to be empirically useful, the postulated connection between available information and expectations must be testable. Note that the concepts of an expectations hypothesis and of the mathematical expectation operator are distinct; in some applications there may be little or no relationship between the two concepts.

Hypotheses about expectations can be divided into two prominent groups. The simpler and less desirable approach postulates an arbitrary, fixed relationship between the expectation of a future outcome and past observations on the same variable. One example is the "myopic" price expectations hypothesis, in which the predicted price in the next period is equal to the current price. More complicated versions of the fixed expectations rule depend on the lag pattern between actual and predicted values. Common alternatives include an arithmetic average of current and past prices; a weighted sum of current and past prices with geometrically declining weights; and various weighted averages where the weights follow more complex frequency distributions.[6] Under each of these alternatives, expectations about the future are adjusted as additional information becomes available. The magnitude and speed of adjustment depends on the number of lags assumed in the expectations formula and on the distribution of the weights.

The second approach, referred to as the "rational expectations" hypothesis, has been developed to address a serious deficiency in fixed expectations rules. The deficiency of fixed expectations rules is that the formulas are not derived from fundamental hypotheses about how prices and costs are determined. Consequently, predicted changes in prices and costs obtained from the formulas may diverge widely and systematically from actual outcomes over time. To address the deficiency, the rational expectations approach assumes that underlying stochastic influences on prices and costs can be represented as stochastic processes which obey known probability distributions. The distributions are assumed to be consistent with the fundamental determinants of market outcomes (specifically, technology, influences on demand, and regulatory constraints).[7] Decision makers use this knowledge to forecast unknown

6. On adaptive expectations, see, for example, Cagan (1956); on more complicated lag distributions see any comprehensive econometrics text, such as Theil (1971, chapter 6).

7. See Sargent (1981) and references cited therein for a formal discussion of rational expectations in linear economic models; see also Maddock and Carter (1982) for an informal review of strengths and weaknesses of the approach.

future influences optimally and to formulate optimal decisions contingent on future random outcomes. Expectations and decision plans are conditional on past realizations of the stochastic processes and are continually updated as new observations become available.

The assumption that firms know the underlying distribution of random "shocks" and use this knowledge for optimal forecasting of future influences on decisions is a strong one. Weakening this assumption, by allowing for phenomena such as Bayesian "learning" about the underlying probability structure, would be highly desirable in a broad range of empirical applications, not just in studies of nonrenewable resource supply. As noted subsequently, however, practical extensions of the methodology encounter serious difficulties.

An important feature of the rational expectations approach is that firms' expected future profits, and hence their optimal current decisions, generally will depend on the variances and higher moments of the underlying probability distribution, not simply on their mathematical expectations. In this situation, the process of forecasting future prices and costs cannot be separated from the process of determining optimal decisions; that is, optimal decisions cannot be derived from the deterministic supply functions in chapter 2 by substituting the mathematical expectations of prices and other influences for their unknown future values.

The theoretical models discussed in this chapter presume rational expectations. Despite the limitations noted below, this approach offers a promising avenue for empirical research, as emphasized in the recommendations in chapter 7 for future investigation. In sharp contrast, almost all empirical studies of resource supply either ignore uncertainty entirely, by assuming perfect foresight, or address uncertainty by using fixed expectations rules such as myopic expectations or weighted averages of past observations.[8] This gap between theory and application raises questions about the interpretation of the empirical results and about the reliability of forecasts based on them.

EFFECTS OF UNCERTAINTY

Uncertainty about future prices and costs can affect supply decisions in three ways. First, if producers update their expectations and decision plans based on new observations of random price and cost outcomes, then unpredictable fluctuations in prices and costs will cause actual

8. The few exceptions to this general characterization of the empirical supply literature are noted in chapter 6.

decisions to deviate from their expected paths.[9] Second, aversion to risk can cause firms to adopt different decision rules, and hence a different path of decisions for any realization of the random price and cost processes, than would be followed in the absence of uncertainty. These influences are discussed in subsequent sections.

The third potential influence, which is the topic of this section, concerns the effect of uncertain price and cost variability on decision rules and expected decisions by risk neutral firms. With risk neutrality, firms seek to maximize the expectation of the present value of net revenues over time, in contrast to the known present value function used in chapter 2. The simplest way to introduce uncertainty in this context is by using the "certainty equivalence" approach, which retains the deterministic supply functions derived earlier but replaces unknown future influences with their expected values (conditional or available information).

The certainty equivalence approach has the advantage that deterministic decision rules continue to hold, since these rules are less complicated than their counterparts under alternative treatments of uncertainty. The only new element required in the theory is a hypothesis about how expectations are formed. However, to use this approach under the hypothesis of risk neutrality and optimal decision making, the certainty equivalence supply functions must be consistent with decision rules that maximize expected net present value. In particular, decisions must depend only on the mathematical expectation of random future influences, and not on the variances and higher moments of these influences.

Much of the discussion that follows is concerned with the conceptual and practical relevance of the certainty equivalence approach. Drawing upon the work of Pindyck (1980, 1981a, 1981c, 1982), the discussion highlights the limitations of certainty equivalence, particularly the implication that random price and cost variability has no effect on behavior. In moving beyond the certainty equivalence approach, however, the models quickly become unwieldly. In practical terms, the certainty equivalence approach is suited to empirical application, but the strong assumptions underlying the approach narrow the scope of the model.

Extraction Decisions

We begin by considering the effects of uncertainty on extraction decisions when there are no additions to reserves, and when the paths of

9. With expectations continually being updated, expected price and cost paths at any moment in time are conditional on current and past observations and are not the same as the paths expected *a priori* at the initial time $t = 0$. Throughout this section, references to expected prices and costs refer to conditional expectations.

future prices and reserves are uncertain. Uncertainty about the price path is introduced by assuming that the difference between the actual percentage change in the resource price and the expected percentage change is a normally distributed random variable with a zero mean.[10] Uncertainty about reserves is modeled by assuming that the actual rate of decline in reserves differs from the expected rate—which is equal to the rate of extraction—by another normally distributed disturbance with zero mean.[11] This disturbance term can be interpreted as unpredictable "revisions" of reserve estimates, which is a matter of some importance in petroleum data (see chapter 6); it can also be interpreted as a proxy for stochastic cost fluctuations. Adopting a rational expectations approach, we assume that the producer knows the probability distributions of the two disturbance terms (conditional on the current price and reserve stock) and uses this information to formulate expectations of the future. The variances of future prices and reserve stocks, conditional on their current magnitudes, increase as the forecasts become more distant.

In this context, it can be shown that the expected rate of decline in the optimal extraction rate over time generally differs from the certainty equivalence decline rate—the rate at which output would decline in the absence of uncertainty about prices and reserves—by an amount that is proportional to the variances of the price and reserve disturbances.[12] This conclusion highlights the limitations of models that use the certainty equivalence approach, although the implications of uncertainty for supply behavior are ambiguous without further assumptions on the cost function.[13]

Concerning the effect of price uncertainty, it further can be shown that if marginal extraction cost is convex in output, so that cost rises at an increasing rate with the rate of extraction for any level of reserves, then

10. See Pindyck (1981a, p. 283). Under this assumption, the true price is lognormally distributed. Uncertainty about prices is expressed in terms of percentage changes, in contrast to absolute rate of change, to avoid the possibility of negative prices when the error term is negative and sufficiently large. Aside from this perverse possibility, however, the results are essentially the same in the two cases.

11. See Pindyck (1980, p. 1207). With this specification, the absurd outcome of negative reserves is possible when the error term has a negative value larger than existing reserves. Pindyck (1981c) explores an alternative specification in terms of percentage changes, but the specification has the troublesome feature that reserves increase over time when extraction is zero even though the model does not allow for explicit additions to reserves through exploration.

12. Proof of these assertions and interpretations of the results can be found in Pindyck (1980, pp. 1210–1212; 1981a, pp. 284–286); see also the appendix to this chapter.

13. Two additional assumptions that affect the conclusions are, first, mutual independence of the error terms for prices and reserves and, second, positive values of the decision variables at every moment in time, including expected decisions at all future dates. The second assumption precludes corner solutions.

the expected rate of decline in output is larger than the certainty equivalence rate (Pindyck, 1981a, pp. 285–286). Uncertainty about prices induces a relative shift in output toward the present and a faster rate of depletion of reserves. In contrast, if marginal extraction cost is concave, so that cost rises at a decreasing rate with output, then price uncertainty pushes output into the future and slows the rate of depletion.

The impact of uncertainty about reserves is more difficult to analyze because of the way changes in reserves affect future as well as current extraction costs. In the special case in which marginal extraction cost is additively separable and linear in the extraction rate, so that marginal cost has the general form $E_q(q, R) = h(R) + \beta q$, reserve uncertainty leads to a faster depletion response if and only if $h'' > 0$. This condition implies that unit extraction cost rises at an increasing rate as reserves decline; in other words, extraction cost is convex in reserves.

A simple intuitive explanation for the influence of price uncertainty becomes apparent when random fluctuations in price offset each other, so that the average price is constant over time. Temporary increases in the price relative to the average will induce short-run increases in output, and downward price fluctuations will induce temporary reductions in output. If marginal extraction cost is increasing at an increasing rate, then the temporary increases in output will raise marginal cost by more than the corresponding reductions in marginal cost when output declines. Therefore, if the firm responds symmetrically to price fluctuations by making equal corresponding changes in output around the average, then the average marginal cost over time will be higher than marginal cost along the mean output trend. The increase in cost provides an incentive to speed depletion, in contrast to the certainty equivalence outcome, by choosing a larger initial extraction rate and a more rapid decline in output (on average) over time. Similar reasoning applies to fluctuations in reserves.[14] If extraction cost rises at an increasing rate as reserves decline, then an unforeseen drop in reserves raises marginal cost by more than the decline in marginal cost from an equal stochastic increase in reserves. Again, firms have an incentive to speed depletion.

The appendix to this chapter discusses related findings under more general specifications of the extraction cost function. We also show there that for the certainty equivalence approach to hold—that is, for the

14. The reasoning here is similar to Pindyck's (1980) analysis of *market* behavior with a less than perfectly elastic market demand for the resource. Pindyck assumes that demand satisfies $p = yf(q)$, where $f(q)$ is the nonstochastic demand component and y is a disturbance, whereas we take the special case in which f is constant. In this way, we focus attention on the behavior of an individual producer whose output decisions do not affect the market price. The derivation of our conclusion is discussed in the appendix to this chapter.

extraction rule to be independent of the variances of price and reserve fluctuations—the extraction cost function must satisfy the conditions

$$E_{qqq} = E_{qqR} = E_{qRR} = 0 \tag{1}$$

where the triple subscripts indicate third derivatives of E. The first two conditions imply that the slope of marginal cost is independent of both the extraction rate and the level of reserves, while the third condition implies that marginal extraction cost is linear in reserves. Note that all of the conditions are satisfied by a linear-quadratic extraction cost function $E(q, R) = (\eta - \gamma R)q + (\beta/2)q^2$, where $\eta > 0$, $\gamma > 0$, and $\beta > 0$ are fixed parameters. This conclusion is important to keep in mind in subsequent discussions of rational expectations models based on a linear-quadratic cost specification, in which the certainty equivalence approach is used.

Exploration, Reserve Development, and Joint Products

Bringing reserve additions into the picture substantially complicates the analyses as always, and relatively few definitive results are available. Pindyck (1980, pp. 1214–1217) considers a case in which exploratory effort is a decision variable and the marginal product of effort—and, hence, the rate of addition to reserves—is subject to random fluctuations for any level of effort. If the marginal product of exploratory effort is convex in the random disturbance term, then the conclusion is analogous to that reported above for the extraction decision. Specifically, the effect of uncertainty is to increase the average marginal product over time by more than the expected value in the certainty equivalence case and to reduce exploratory effort, though effort will also decline less rapidly over time. The opposite conclusion follows if the marginal product of exploratory effort is concave in the disturbance term, while the uncertainty has no effect on the expected dynamics of exploration if the marginal product is linear in the disturbance.

Further insight into the effects of uncertainty on exploration and development can be gained by examining a stochastic analogue of the reserve additions model discussed in chapter 2. Reserve uncertainty is brought into the model by assuming that expected additions to reserves are proportional to effort and differ from actual additions by a random disturbance. In the appendix, we show that uncertainty about additions to reserves generally has ambiguous effects on both extraction and reserve additions. However, in the special case of a marginal cost function that is additively separable in reserves and in the extraction rate and linear in the extraction rate, uncertainty about additions to reserves produces more rapid depletion if and only if unit costs rise at an increasing rate with depletion. A similar line of argument can be applied to exploration and development effort. We also show that conditions analogous to equation (1) are required for the certainty equivalence approach to hold for both extraction and reserve additions, so that a linear-quadratic specification of development cost is needed to use the approach.

The situation is even more complex in the presence of joint products.[15] In addition to the issues raised above, responses to uncertainty also depend on the correlations among the prices of the joint products and on the cross-price supply elasticities of these outputs. However, in the special case of additively separable marginal production costs that are linear in output rates, it can be shown that uncertainty about reserves leads to more rapid depletion if correlations among output prices are sufficiently small.

"Learning" Responses

The discussion of uncertainty about reserves to this point has treated the variance of fluctuations in reserves as independent of producer decisions. However, it is perhaps more reasonable to suppose that uncertainty about reserves falls as cumulative extraction increases and the stock of remaining reserves declines. In the appendix, we show that this learning effect compensates for the cost associated with reserve fluctuations and thus counters the tendency for producers facing uncertainty about reserves to hasten depletion. However, this conclusion depends on the assumption of risk neutrality, as shown subsequently.

The presence of uncertainty also provides an incentive for exploration and development designed to reduce uncertainty as well as to add new reserves.[16] In fact, information obtained from exploration can be valuable enough to lead to rising exploratory effort over time, even though cost also is rising.[17] This learning influence also works counter to the tendency toward more rapid depletion. In short, when the information value of exploration and extraction is brought into the analysis, the net

15. Pindyck (1982) presents a joint products model that includes uncertainty about demand and price but does not examine uncertainty about reserves.

16. Recall from the discussion of externalities in chapter 3 that information—which can be interpreted as a lower variance of reserve estimates—can be viewed as a joint product (with actual discoveries) of exploratory effort.

17. See Pindyck (1980, pp. 1212–1215) for an example. For related analyses, see Arrow and Chang (1982) and the discussion and citations in Dasgupta and Heal (1979, chapter 13).

effect of uncertainty on extraction and reserve additions becomes even less clear-cut.

RATIONAL EXPECTATIONS IN LINEAR SUPPLY MODELS

The analysis of supply behavior under uncertainty is considerably simplified if the fluctuations in marginal cost are assumed to be linear and the stochastic disturbances related to uncertainty are assumed to affect price and cost additively. These conditions, together with certain assumptions of stability in the random price and cost processes described below, make it possible to derive explicit formulas for the supply functions and for expectations. It is not surprising, therefore, that rational expectations models of supply behavior frequently are couched in these terms.

The linear rational expectations approach has been used primarily in macroeconomic models but more recently has been applied to nonrenewable resource supply.[18] These applications are based upon four fundamental sets of assumptions. The first is that costs satisfy a linear-quadratic specification (linear in stock variables, quadratic in decision variables), so that marginal costs are linear, and that random fluctuations affect prices and costs in a linear fashion. The second assumption is that stochastic price and cost processes can be represented as autoregressions; that is, as weighted sums of past realizations plus a purely random disturbance.[19] At the level of the individual firm, these processes are assumed to be exogenous, while at the market level they are the product of more fundamental exogenous influences along with interactions between demand and supply, as discussed in chapter 5. The weights in these representations are not constrained to follow a particular distributional pattern but are required to satisfy stationarity conditions that ensure that prices and costs do not "explode." In addition, the lag weight distributions are assumed to be consistent with the fundamental determinants of price and cost in the market. Prior restrictions also are placed upon the structural parameters of the cost functions to rule out explosive output behavior along the expected price and cost paths.[20]

18. See Lucas (1976), Sargent (1979, 1981), and Maddock and Carter (1982) for discussions of the linear rational expectations approach in macroeconomics. Recent applications of this approach to petroleum supply include Epple and Hansen (1979, 1981); Epple, Hansen, and Roberds (1983); and Epple (1983). Roberds (1982) applies the approach to mineral industries such as nickel under the assumption of noncompetitive suppliers.

19. Granger and Newbold (1977, chapter 1) contains a general discussion of autoregressive and other stochastic processes.

20. Granger and Newbold (1977, chapter 1) and Hansen and Sargent (1980) contain more precise descriptions of these conditions.

The third assumption is that producers use knowledge of these autoregressive processes, along with knowledge of past and current realizations, to form rational expectations of prices and costs. Rational expectations in this context are interpreted to be forecasts that minimize mean-square forecast error.[21] Finally, it is assumed that in each period current decisions are not the "corner solutions" of zero activity, *and* that corner outcomes are not expected at any future date. In other words, nonnegativity constraints on decisions are never binding. Capacity constraints also are ignored.

The assumptions of linear-quadratic cost specifications and absence of binding nonnegativity or capacity constraints mean that uncertainty has no effect on the *expected* dynamics of supply decisions and that the certainty equivalence methodology is appropriate, as demonstrated above. Under these conditions, the variances of the random disturbances do not affect the form and content of decision *rules*, but will cause actual decisions to deviate from anticipated behavior because of price and cost fluctuations that influence both decisions and expectations. This feature represents both a strength and a weakness of the assumptions. On the one hand, it substantially simplifies the specification and estimation of supply functions. However, under a certainty equivalence approach with additive stochastic processes, there is no scope for the influence of random price and cost variances when marginal costs are nonlinear, for declining uncertainty about reserves as reserves are depleted, or for the learning effect of exploration in reducing the variance of reserve estimates. The assumption that nonnegativity and capacity constraints can be ignored also is crucial to the applicability of the certainty equivalence formulation, as indicated in the next section.

The assumption that producers have full knowledge of the stochastic price and cost processes (lag weights and distributions) is obviously strong. However, this assumption cannot be relaxed—in favor of a Bayesian approach in which the producer "learns" about the probability distributions—and still retain analytical and computational tractability.[22] The assumption also proves useful in deriving consistent estimates of supply parameters, and in separating interactions between price and supply at the market level, as indicated in chapter 5.

21. Sargent (1979, chapter 10). If errors are normally distributed, the minimization applies to all possible forecasting formulas; if they are nonnormal, the weaker but computationally tractable criterion of minimization over all linear forecasting formulas is used. See also Karlin and Taylor (1975, chapter 9).

22. Note that "learning" has a different interpretation in this context than in the previous paragraph, where the producer was assumed to have full knowledge of the probability distributions but the variances of the distribution were assumed to be functionally related to the producer's decisions.

Optimal Extraction Behavior

To illustrate how the linear rational expectations approach has been employed, we draw upon the work of Epple and Hansen (1979, 1981) concerning optimal extraction decisions. The cost function is assumed to obey the general form

$$E(q_t, R_t) = (\eta_t - \gamma R_t)q_t + (\beta/2)q_t^2 \tag{2}$$

where $\gamma > 0$ and $\beta > 0$ are fixed parameters and η_t is a random variable.[23] This random cost shock is assumed to be independent of reserves, and reserves are assumed to be known. The implications of reserve uncertainty are discussed below.

The price of extracted output p_t is determined exogenously to the extraction decision and is assumed to follow an autoregressive process

$$p_t = \phi_0 + \phi_1 p_{t-1} + \phi_2 p_{t-2} + \cdots + \phi_N p_{t-N} + e_t^p \tag{3}$$

while the cost disturbance is also exogenous and follows an autoregressive process

$$\eta_t = \psi_0 + \psi_1 \eta_{t-1} + \psi_2 \eta_{t-2} + \cdots + \psi_M \eta_{t-M} + e_t^\eta \tag{4}$$

In equations (3) and (4), the terms e_t^p and e_t^η represent serially and mutually uncorrelated normal random variables having zero means. The producer is assumed to know the lag weights $\{\phi_i, i = 0, \ldots, N\}$ and $\{\psi_i, i = 0, \ldots, M\}$ and, at each time t, the current and past realizations of p and η.[24]

Using the identity $R_{t+1} = R_t - q_t$, net extraction revenue can be written as

$$\pi_t(R_t, R_{t+1}) = [p_t - (\eta_t - \gamma R_t)](R_t - R_{t+1}) - (\beta/2)(R_t - R_{t+1})^2 \tag{5}$$

Differentiating equation (5), we obtain

$$\partial\pi_t/\partial R_t = p_t - \eta_t + (2\gamma - \beta)R_t - (\gamma - \beta)R_{t+1} \tag{6}$$

$$\partial\pi_t/\partial R_{t+1} = -(p_t - \eta_t) - (\gamma - \beta)R_t - \beta R_{t+1} \tag{7}$$

Notice that these derivatives are linear in R_t, R_{t+1}, and in the random variables p_t and η_t. This property plays a crucial role in the derivation of the optimal extraction path.

23. We have changed Epple and Hansen's notation to conform more closely with previous usage.

24. As noted before, the assumption of normality can be relaxed if the expectations formulas used by the producer are assumed to be linear in current and past realizations. Epple and Hansen consider a slightly more general setting in which p_t and η_t are elements of larger vector autoregressive processes. The other components are interpreted to be variables that the firm believes are correlated with p_t and η_t, and thus help to predict price and cost shocks.

Using equations (6) and (7) and the solution techniques described in Hansen and Sargent (1980), it can be shown that the optimal output plan satisfies a linear equation of the form

$$q_t = A + CR_t + v_t + u_t \tag{8}$$

where A and C are constants that depend on the underlying cost parameters and the discount factor d. The terms u_t and v_t are given by

$$v_t = \sum_{j=0}^{\infty} B^j \mathscr{E}_t\left(p_{t+j} - dp_{t+j+1}\right) \tag{9}$$

$$u_t = \sum_{j=0}^{\infty} B^j \mathscr{E}_t\left(\eta_{t+j} - d\eta_{t+j+1}\right) \tag{10}$$

where B is a constant that depends on the underlying parameters; d is the discount factor; $\mathscr{E}_t(p_{t+j} - dp_{t+j+1})$ is the mathematical expectation of $p_{t+j} - dp_{t+j+1}$ given current and past prices at time t; and $\mathscr{E}_t(\eta_{t+j} - d\eta_{t+j+1})$ is defined similarly.

Equations (8) through (10) imply that optimal output is a linear function of current reserves, expected future price appreciation, and expected future appreciation in the stochastic cost term η_t. This is a special case of the general supply functions described in chapter 2. Moreover, it can be shown that this supply function is exactly the same as its certainty equivalent: if we replace unknown future prices and cost shocks with conditional expectations, and solve the resulting deterministic present-value-maximization problem, we obtain an identical set of expressions describing optimal extraction.

An important virtue of this approach is that equations (8) through (10) can be translated into a formula for extraction in terms of current and past prices and cost shocks, under the assumption that expectations are formed rationally. Using prediction formulas developed in Hansen and Sargent (1980), it can be shown that the conditional expectations of prices, $\mathscr{E}_t p_{t+j}$, are given recursively by[25]

$$\begin{aligned}
\mathscr{E}_t p_{t+1} &= \phi_0 + \phi_1 p_t + \phi_2 p_{t-1} + \cdots + \phi_N p_{t-N+1} \\
\mathscr{E}_t p_{t+2} &= \phi_0 + \phi_1 \mathscr{E}_t p_{t+1} + \phi_2 p_t + \cdots + \phi_N p_{t-N+2} \\
&\ \ \vdots \qquad \vdots \qquad \vdots \qquad \vdots \qquad \vdots
\end{aligned} \tag{11}$$

where the ϕ_i terms are the autoregressive weights in equation (3). Similarly, expectations of future cost shocks are given by

$$\begin{aligned}
\mathscr{E}_t \eta_{t+1} &= \psi_0 + \psi_1 \eta_t + \psi_2 \eta_{t-1} + \cdots + \psi_M \eta_{t-M+1} \\
\mathscr{E}_t \eta_{t+2} &= \psi_0 + \psi_1 \mathscr{E}_t \eta_{t+1} + \psi_2 \eta_t + \cdots + \psi_M \eta_{t-M+2} \\
&\ \ \vdots \qquad \vdots \qquad \vdots \qquad \vdots \qquad \vdots
\end{aligned} \tag{12}$$

25. Also refer to Granger and Newbold (1977, chapter 4).

These conditional expectations also represent optimal forecasts in the sense of minimum mean-square forecast error.[26]

Substituting equations (11) and (12) into equations (9) and (10), and then substituting the resulting expressions for u_t and v_t in equation (8) yields an output supply function that is linear in current reserves, in a weighted sum of current and past prices, and in a weighted sum of current and past cost disturbances η.[27] Linearity of the supply function in the observable explanatory variables $(R_t, p_t, p_{t-1}, \ldots)$ clearly is an advantage for empirical applications. Linear regression estimates of the coefficients of the variables in the supply function may be used to derive estimates of the underlying parameters in equations (2) through (4). Moreover, the linear dependence of supply on a weighted sum of current and past realizations of the cost shock provides a theoretical rationale for the stochastic disturbance term in an econometric supply equation. The error term arises under the assumption that the econometric analyst (unlike the resource producer) cannot observe the cost shock.[28]

A difficulty arises because the unobservable disturbance in the supply function will be serially correlated unless the underlying cost shock is serially uncorrelated (that is, $\psi_i = 0$ in equation (4) for all $i \geq 1$). This result has important implications for empirical analysis, since equation (8) also contains a lagged dependent variable: reserves R_t depend on the previous period's extraction q_{t-1} as well as on extraction in earlier periods. In this situation, ordinary least squares estimates of the supply function parameters in equation (8) will be inconsistent. It follows that derived estimates of the underlying parameters in equations (2) through (4) will be inconsistent as well.

Extensions of the Rational Expectations Approach

The rational expectations approach just described can be applied to the case in which estimates about reserves fluctuate stochastically.[29] Suppose that reserves satisfy

$$R_{t+1} = R_t - q_t + \lambda_t \tag{13}$$

26. The proviso noted earlier applies again if p and η are not normally distributed.

27. Epple and Hansen (1981, pp. 109–112) illustrate this procedure for alternative special cases of the stochastic price and cost processes given by equations (3) and (4). The alternative price processes considered include a random walk, stochastic fluctuation around a constant price, and convergence of price to some asymptotic value over time; their cost shock examples are "white noise" (serially uncorrelated disturbances) and a first-order autoregression.

28. See Hansen and Sargent (1980) for a discussion of this issue under general specifications of the stochastic shock processes.

29. We are not aware of any previous effort in the literature to address this question, despite the apparent importance of uncertainty about reserve estimates (see chapter 6).

where λ_t is a random variable. Net revenue can be written as

$$\pi_t(R_t, R_{t+1}) = [p_t - (\eta_t - \gamma R_t)](R_t - R_{t+1} + \lambda_t) - (\beta/2)(R_t - R_{t+1} + \lambda_t)^2 \qquad (14)$$

which is analogous to equation (5). Differentiation of equation (14) yields

$$\partial\pi_t/\partial R_t = p_t - \eta_t - (\beta - \gamma)\lambda_t + (2\gamma - \beta)R_t - (\gamma - \beta)R_{t+1} \quad (15)$$

$$\partial\pi_t/\partial R_{t+1} = -(p_t - \eta_t - \beta\lambda_t) - (\gamma - \beta)R_t - \beta R_{t+1} \qquad (16)$$

Like equations (6) and (7), equations (15) and (16) are linear in (R_t, R_{t+1}) and in the random variables (p_t, η_t, λ_t). Therefore, the general approach described above remains applicable in this context. The only potential difference arises from the possible covariance between reserve and cost uncertainty, in which case the expectation formulas analogous to equation (12) for $\mathscr{E}_t\eta_{t+j}$ and $\mathscr{E}_t\lambda_{t+j}$ must be modified to account for the covariance between η and λ.

The linear rational expectations model can be further extended to include both extraction and reserve additions decisions by an integrated firm. Suppose that the cost of additions to reserves satisfies

$$C(y_t, z_t) = (\xi_t + \delta z_t)y_t + (\omega/2)y_t^2 \qquad (17)$$

where ξ_t is a random cost shock analogous to η_t in equation (2) and $\delta > 0$ and $\omega > 0$ are fixed parameters. Linear equations for both optimal extraction and reserve additions then can be derived using a vector analogue of the approach taken by Epple and Hansen.[30]

Vector linear models with rational expectations also can be used to describe market-level supply behavior and interactions among supply, demand, and institutional constraints, as indicated in the next chapter. In this setting, expectations of future prices in the supply functions are based on expectations of exogenous influences on demand (for example, shifts in income, wealth, or taste), which in turn influence market equilibrium prices. This approach is particularly useful in sorting out market interactions and identifying supply responses using market data.

CORNER SOLUTIONS AND CERTAINTY EQUIVALENCE BIAS

In discussing applications of the certainty equivalence approach, we have noted the importance of the assumption that nonnegativity constraints on decisions are not binding; in other words, that producers have no incentive to temporarily delay extraction or additions to reserves. Pindyck (1981a, pp. 286–287) uses a simple model of extraction with extraction cost independent of reserves to show that uncertainty about prices slows

30. See Hansen and Sargent (1981) for a general discussion of the vector approach.

down depletion relative to the certainty equivalence outcome when the possibility of delaying extraction is taken into account. This conclusion counters the tendency toward more rapid depletion when nonnegativity constraints are ignored and indicates that the net effect of uncertainty on behavior may be ambiguous, even when specifications of cost functions give determinate findings in the certainty equivalence case.

The balance of this section uses a simple two-period model to illustrate the problems posed by binding constraints in the linear rational-expectations approach. The model also illustrates how the formation of expectations affects decision making under uncertainty, and specifically how uncertainty affects the form and content of decision rules.

Suppose that the extraction cost function satisfies equation (2) with $\eta > 0$ a fixed parameter, and a known initial reserve stock; thus there is no uncertainty about costs and reserves. Let p_0 and R_0 be the price of extracted output and the stock of reserves in the initial period ($t = 0$). Both of these variables are assumed to be known. Let $R_1 = R_0 - q_0$ be the stock of reserves remaining at the start of the second period ($t = 1$). The second period price p_1 is unknown but obeys a known probability density function $\phi(p_1 | p_0)$, which may depend on p_0. We assume that $\eta - \gamma R_0 > 0$, so that marginal extraction cost is positive in view of equation (2).

Using the backward recursion approach described in chapter 2, we first determine the second-period output rate $q_1 = f(p_1, R_1)$ that maximizes second-period profits, given any realization of the price p_1 and remaining reserves of R_1. Using equation (2) and allowing for the corner solution $q_1 = 0$, the function $f(p_1, R_1)$ can be shown to satisfy

$$f(p_1, R_1) = \begin{cases} (p_1 - \eta + \gamma R_1)/\beta, & p_1 - \eta + \gamma R_1 \geq 0 \\ 0, & p_1 - \eta + \gamma R_1 < 0 \end{cases} \tag{18}$$

Maximum second-period profits are given by

$$\begin{aligned} \pi_1(p_1, R_1) &= \pi_1(p_1, R_0 - q_0) \\ &= (p_1 - \eta + \gamma R_1) f(p_1, R_1) - (\beta/2)[f(p_1, R_1)]^2 \end{aligned} \tag{19}$$

Notice that $\pi_1 = 0$ in equation (19) if $f = 0$.

Next we determine the initial output rate that maximizes the expected discounted sum of profits, given that second-period output is chosen optimally according to equation (18). This expected sum is given by

$$\begin{aligned} V(q_0, p_0, R_0) &= p_0 q_0 - E(q_0, R_0) \\ &\quad + d \int_{\eta - \gamma R_1}^{\infty} \pi_1(p_1, R_0 - q_0) \phi(p_1 | p_0)\, dp_1 \end{aligned} \tag{20}$$

where $R_1 = R_0 - q_0$, $E(q_0, R_0)$ is as in equation (2), π_1 is as in equation (19), and d is the discount factor. The integral expression in equation (20) is the expected value of second-period profit, given reserves of $R_1 = R_0 - q_0$ and given that second-period extraction is chosen optimally. Note that the lower range of integration is truncated at $\eta - \gamma R_1 = \eta - \gamma R_0 + \gamma q_0 \geq \eta - \gamma R_0 > 0$, because for $p_1 \leq \eta - \gamma R_1$, second-period output and profit are zero as noted above.

Differentiating equation (20) with respect to q_0 and using equation (18), it can be shown that

$$\begin{aligned}\partial V/\partial q_0 &= p_0 - \eta + \gamma R_0 - \beta q_0 + d\int_{\eta-\gamma R_1}^{\infty} E_R[f(p_1, R_1), R_1]\phi(p_1|p_0)\,dp_1 \\ &= p_0 - \eta + \gamma R_0 - \beta q_0 \\ &\quad -(\gamma d/\beta)\int_{\eta-\gamma R_1}^{\infty}[p_1 - \eta + \gamma(R_0 - q_0)]\phi(p_1|p_0)\,dp_1 \qquad (21)\end{aligned}$$

The integral expression in equation (21) is the expected user cost of current extraction and is analogous to the user cost term in equation (9) of chapter 2. Allowing for the possibility of zero current extraction (if the initial price is sufficiently low relative to the expected future price), the marginal profit rule for optimal q_0 is $\partial V/\partial q_0 \leq 0$, $q_0 \partial V/\partial q_0 = 0$.

We now compare this decision rule to that derived using a certainty equivalence approach. In that situation, the producer assumes that the second-period price will be its mathematical expectation

$$\bar{p}_1 = \bar{p}_1(p_0) = \int_0^{\infty} p_1\phi(p_1|p_0)\,dp_1 \qquad (22)$$

conditional on the current price p_0. Optimal second-period output, given q_0 and this expectation, satisfies $\bar{q}_1 = f(\bar{p}_1, R_1)$ where f is as expressed in equation (18). Notice that we allow for the possibility of zero expected output in the second period if the expected second-period price is sufficiently small relative to the first-period price.

In the certainty equivalence model, expected discounted total profit is given by

$$\bar{V}(q_0, p_0, R_0) = p_0 q_0 - E(q_0, R_0) + d\pi_1(\bar{p}_1, R_0 - q_0) \qquad (23)$$

where π_1 is as in equation (19) with second-period output of $f(\bar{p}_1, R_1)$. Differentiating equation (23) yields

$$\begin{aligned}\partial \bar{V}/\partial q_0 &= p_0 - \eta + \gamma R_0 - \beta q_0 + dE_R[f(\bar{p}_1, R_1), R_1] \\ &= p_0 - \eta + \gamma R_0 - \beta q_0 \\ &\quad - \begin{cases}(\gamma d/\beta)[\bar{p}_1 - \eta + \gamma(R_0 - q_0)], & \bar{p}_1 \geq \eta - \gamma(R_0 - q_0) \\ 0, & \bar{p}_1 < \eta - \gamma(R_0 - q_0)\end{cases} \qquad (24)\end{aligned}$$

Comparing equations (21) and (24), it can be shown that $\partial V/\partial q_0 < \partial \bar{V}/\partial q_0$ for all $q_0 \geq 0$, provided that the probability of $p_1 \leq \eta - \gamma(R_0 - q_0)$ is greater than zero. Thus, if $\text{Prob}\{p_1 \leq \eta - \gamma R_0\} > 0$, then the optimal output rate derived from equation (21) generally will be less than the certainty equivalence rate derived from equation (24).

Two important points can be gleaned from this discussion. First, application of the certainty equivalence approach when nonnegativity constraints are binding generally leads to biased conclusions about optimal supply decisions. In the two-period example, the certainty equivalent output rate in the initial period is biased upward. Although this result does not readily extend to longer planning horizons, it does draw attention to the possibility of specification error using the certainty equivalence approach.[31]

This problem can be avoided only by placing additional (and possibly overly stringent) lower bounds on the distribution of future prices to ensure that nonnegativity constraints are never binding in future periods. If these restrictions do not hold, then the optimal output rule depends on the entire distribution of prices, not just the mean, as illustrated by equation (21).[32]

The other observation is that when nonnegativity constraints are either currently binding or potentially binding in the future, even the certainty equivalence decision rule generally is more complex than the linear rules presented in the previous section, which were based on the assumption of nonbinding constraints. This can be seen by noting that the marginal profit rule $\partial\overline{V}/\partial q_0 \leq 0$, $q_0 \partial\overline{V}/\partial q_0 = 0$ for certainty equivalence output is a complex nonlinear function that allows for either zero extraction in the first period or the possibility of zero extraction in the second period. The possibility of binding capacity constraints on extraction only aggravates these nonlinearities, which also carry over to exploration and development decisions. Specifically, the possibility that additions to reserves cease before extraction begins introduces an additional source of nonlinearity.

It follows that application of the certainty equivalence approach to nonrenewable resource supply introduces several potential sources of specification error if corner solutions are prevalent. In addition to those noted above, there is the question of whether linear-quadratic cost specifications such as equation (2) are adequate representations of actual resource supply costs. While the linear decision rules derived from these specifications are substantially more tractable than those sketched in this section, there is an important tradeoff between practicality—ignoring the

31. This bias is analogous to the "irreversibility effect" discussed in Henry (1974) and Dasgupta and Heal (1979, p. 397–400).

32. This caveat also applies to Pindyck's analyses of nonlinear rational expectations models: his finding that decisions depend only on the means and variances of random disturbances (versus their entire distributions) depends on the assumption of nonbinding nonnegativity constraints.

various nonlinearities discussed in this section—and the problem of misspecification.

CONCLUDING REMARKS

The theme of this chapter is that uncertainty about future prices and costs, and producer responses to uncertainty, introduce a new dimension to the analysis of resource supply behavior. Along with the dynamic interactions among decisions and influences highlighted in chapters 2 and 3, two additional concerns must be addressed: how unpredictable fluctuations in prices and costs can cause deviations of decisions from their expected paths, as producers revise both their actions and expectations, and how supply functions or decision rules themselves are altered by these fluctuations.

Because expectations are not observable, a separate hypothesis about expectations (in addition to a hypothesis about behavior rules for given expectations) is needed to address these issues. Most of this chapter has been concerned with the analysis of behavior under the assumption of "rational" expectations because this is the only tractable approach capable of addressing the issues. The focus on rational expectations, in turn, has impelled us to emphasize its limitations as well as its strengths.

An examination of theoretical resource supply models based upon rational expectations indicates that, in general, decisions depend in a complex, nonlinear way upon the variances and covariances of random price, cost, and reserve fluctuations, as well as on other characteristics of their distributions. In particular, uncertainty about future prices and costs tends to induce more rapid depletion of resources relative to a certainty equivalence outcome in which prices and costs follow their expected paths. However, there are also influences that run counter to this trend, such as the effect of exploration in reducing uncertainty and the possibility of delaying production and reserve decisions when resource prices fall temporarily.

Linear optimal decision rules with the certainty equivalence property can be derived from rational expectations models that assume linear-quadratic cost functions and the absence of binding constraints on decisions. However, the hypothesis that decisions depend only on the mathematical expectations of prices and cost, and not on the variances of price and cost disturbances, represents a limitation of the approach. Consequently, the analytical simplicity and computational tractability of linear certainty equivalence models does not come free. Moreover, while the rational expectations hypothesis assumes that firms use knowledge of stochastic price and cost processes to forecast future values, the specifications of the processes are *ad hoc*. The analysis could be enhanced by deriving specifications of reserve and cost processes from the sampling theories described in Kaufman (1963) and subsequent extensions such as Ramsey (1980).[33]

Another conceptual issue is found in the treatment of decision makers' attitudes toward risk. Throughout this chapter we have focused on risk neutral producers. Fisher (1982, pp. 46–49) points out that aversion to risk generally alters supply responses, but the qualitative effect of risk aversion depends critically on the type of uncertainty under consideration.[34] Pindyck (1980, 1982) analyzes risk averse supply behavior under the assumption that producers maximize the expected discounted sum of "profit utilities" over time, where the curvature of the "utility of profit" function reflects the degree of risk aversion.[35] Pindyck finds that the effect of uncertainty about future prices (or, in a market context, about resource demand) depends on how uncertainty affects prices (linearly versus nonlinearly) and on the curvature of the profit utility function in the producer's objective function. In the special case in which marginal extraction cost is constant, reserve uncertainty leads to more rapid depletion. Whether this conclusion extends to more general cost specifications that include depletion effects is unclear. In any event, the effects of risk aversion on behavior generally are far more complex than can be captured by incorporating a simple upward "risk premium" adjustment in the discount rate.[36]

Still another conceptual issue concerns the assumption that producers know the structure of underlying stochastic price and cost processes. In practice, producers presumably have (at most) rough estimates of these distributions, which become modified on the basis of accumulated experience. Relaxing the assumption of full knowledge and allowing for learning about probability distributions would constitute a major advance in the theory. Extensions that incorporate expectations of exogenous changes in these underlying distributions because of shifting market conditions or

33. See also references in Peterson and Fisher (1977, p. 677). Eckbo, Jacoby, and Smith (1978) apply Kaufman's process approach to North Sea oil supply.

34. To illustrate his point, Fisher examines two scenarios: randomly fluctuating prices around a fixed level, and an abrupt decline in the value of reserve holdings due to expropriation or introduction of substitute goods. In the first scenario, with declining output over time, the variance of profit also falls; thus risk averse firms are induced to shift extraction toward the future. In the second case, exposure to risk is minimized by hastening depletion.

35. This approach is based upon the static theory of consumer choice under risk (Arrow, 1971). As noted by Zabel (1971), however, a more direct analogue of static consumer theory is the assumption that producers seek to maximize the utility of expected discounted net present value, rather than the expected discounted sum of single-period profit utilities.

36. See also Dasgupta and Heal (1974, pp. 18–25).

regulatory constraints would be especially useful.[37] As noted previously, some theoretical efforts in this direction have been undertaken.[38] However, as Sargent (1981) argues, incorporating these extensions into empirical analysis of supply behavior (for nonrenewable resources or other commodities) is a very difficult task.

APPENDIX

The material in this appendix describes the effects of uncertainty on extraction and additions to reserves, using a stochastic analogue of the Peterson (1978) model discussed in chapter 2. Formal proofs are omitted; the method of analysis is the same as in Pindyck (1980, 1981a), to which the reader is referred for further discussions of the approach.

With the following exceptions, all notation is the same as that used in chapter 2. Let dz represent the change in cumulative reserve additions over time. With no uncertainty, $dz = y\,dt$, analogous to equation (24) in chapter 2, where $y \geq 0$ represents the rate of reserve addition by the firm. Here we assume that

$$dz = kw\,dt + \sigma_1\,de_1 = y\,dt + \sigma_1\,de_1, \quad y = kw \tag{25}$$

where the decision variable $w \geq 0$ is a measure of exploration and development effort and $k > 0$ is a fixed parameter. In equation (25), the term de_1 represents a stochastic "white noise" with zero mean and unit variance; $y = kw$ is the expected rate of reserve additions, which we assume to be directly proportional to exploratory and development effort. Thus, equation (25) says that expected additions to reserves are directly proportional to effort and differ from actual additions by a random component with zero mean and standard deviation σ_1.

Using equation (25), net changes in reserves, dR, can be written as

$$\begin{aligned} dR &= -q\,dt + dz \\ &= (kw - q)\,dt + \sigma_1\,de_1 \\ &= (y - q)\,dt + \sigma_1\,de_1 \end{aligned} \tag{26}$$

This equation is the continuous-time stochastic analogue of equation (23) in chapter 2, as can be seen by setting $\sigma_1 = 0$ so that there is no uncertainty in additions to reserves. If $y = 0$, then equation (26) reduces

37. As it stands, the theory addresses only two extremes: perfectly anticipated changes in stochastic processes that are taken into account in current decisions, and completely unanticipated changes to which producers promptly respond (after the fact) by modifying decision rules. See chapter 5 for a further discussion of this issue.

38. See, for example, Grossman, Kihlstrom, and Mirman (1977) for a theoretical analysis of learning by doing.

to Pindyck's (1980) model of extraction with random fluctuations in reserves.

The price of extracted output is assumed to follow the stochastic growth path

$$dp = \alpha p\, dt + p\sigma_2\, de_2 \tag{27}$$

This equation says that the expected percentage growth rate in price is α, and the actual percentage growth rate differs from α by a stochastic white noise $\sigma_2\, de_2$ with zero mean and standard deviation σ_2. We assume, as does Pindyck, that stochastic price fluctuations are uncorrelated with random changes in reserves.

The cost of exploration and development, C, is assumed to be a function of effort w and cumulative discoveries z:

$$C = C(w, z) = C(y/k, z) \tag{28}$$

Under the assumption that expected additions to reserves are directly proportional to effort, this cost can also be represented as a function of y and z, as in chapter 2. In either case, the dependence of cost on z reflects the influence of depletion on exploration and development. To simplify notation, we will assume that $k = 1$. This involves no loss of generality under the assumption of direct proportionality, since units of the effort variable w can be rescaled to achieve a proportionality constant of unity.

Now let $\mathscr{E}_t\, dq/dt$ and $\mathscr{E}_t\, dw/dt$ be the expected rate of change in extraction and effort per unit time, where $\mathscr{E}_t$ represents the mathematical expectation of the random changes dq and dw conditional on the magnitudes of R_t, z_t, and p_t at time t. Note that $y = w$ implies $\mathscr{E}_t\, dy/dt = \mathscr{E}_t\, dw/dt$, so that the expected rate of change in effort also represents the expected rate of change in additions to reserves. Furthermore, let $d\bar{q}/dt$ and $d\bar{w}/dt$ represent the changes in output and effort along a certainty equivalence path; in other words, the changes in output and effort at time t, given that prices and addition to reserves follow their expected paths, so that $\sigma_1 = \sigma_2 = 0$ in equations (25) through (27) for all times $t' > t$.

Using Pindyck's solution technique for these expected dynamics (Pindyck, 1980, appendixes A and C), it can be shown that the rates of change satisfy

$$2E_{qq}(\mathscr{E}_t\, dq/dt - d\bar{q}/dt) = -\left(\gamma_1\sigma_1^2 + E_{qqq}q_p^2p^2\sigma_2^2\right) \tag{29}$$

$$\gamma_1 = E_{qqq}\left(q_R^2 + q_z^2 + 2q_Rq_z\right) + E_{qRR} + 2E_{qqR}(q_R + q_z) \tag{30}$$

$$2C_{ww}(\mathscr{E}_t\, dw/dt - d\bar{w}/dt) = -\left(\gamma_2\sigma_1^2 + C_{www}w_p^2p^2\sigma_2^2\right) \tag{31}$$

$$\gamma_2 = C_{www}\left(w_R^2 + w_z^2 + 2w_Rw_z\right) + C_{wzz} + 2C_{wwz}(w_R + w_z) \tag{32}$$

In each of these equations, double and triple subscripts indicate second and third derivatives of cost functions with respect to the indicated arguments. The terms q_p, q_R, and q_z represent the instantaneous rate of change in extraction with respect to the indicated variable along an optimal path; q_p^2, q_R^2, and q_z^2 are the squares of these rates of change. Terms such as w_R or w_p^2 have a similar interpretation with respect to exploration and development effort. We assume that marginal extraction and effort costs are strictly increasing functions of the decision variables, so that $E_{qq} > 0$ and $C_{ww} > 0$. Therefore, the expected rate of change in output or effort is larger (smaller) than the certainty equivalent change if the term in parentheses on the right-hand side of equation (29) or (31) is less than (greater than) zero.

Turning first to uncertainty about the price, suppose that $\sigma_1 = 0$, so that there is no uncertainty about additions to reserves. Then it follows from equation (29) that $\mathscr{E}_t\, dq/dt < d\bar{q}/dt$—expected output declines more rapidly (or increases more slowly) than the certainty-equivalent—if and only if $E_{qqq} > 0$. This condition states that marginal extraction cost is a strictly convex function of the output rate, so that marginal cost rises with q at an increasing rate. Similarly, $\mathscr{E}_t\, dw/dt < d\bar{w}/dt$ if and only if $C_{www} > 0$, so that marginal effort cost is strictly convex in w.

The effect of uncertainty about reserves depends on the signs of γ_1 and γ_2. In general, these signs are ambiguous. In equation (30), for example, it may be supposed that $q_R > 0$ but $q_z < 0$. Moreover, it may be reasonable to suppose that $E_{qRR} > 0$ (marginal cost rises at an increasing rate as reserves decline) but that $E_{qqR} < 0$ (marginal extraction cost becomes steeper as reserves decline). Consequently, the sign of γ_1 generally is ambiguous.

There is, however, an important case in which these ambiguities do not arise. Suppose that E and C satisfy

$$E(q, R) = qh(R) + (\beta/2)q^2, \qquad h' < 0, h'' > 0, \qquad \beta > 0 \tag{33}$$

$$C(w, z) = wm(z) + (\lambda/2)w^2, \qquad m' > 0, m'' > 0, \qquad \lambda > 0 \tag{34}$$

Note that with these specifications, depletion raises the intercept of marginal cost (at an increasing rate) but does not affect the slope. It follows from equation (33) that $E_{qqq} = E_{qqR} = 0$ and $E_{qRR} = h'' > 0$. Consequently, $\gamma_1 = E_{qRR} = h'' > 0$, and reserve uncertainty causes a more rapid expected decline in output than the certainty-equivalent. Similarly, equation (34) implies that $C_{www} = C_{wwz} = 0$ and $C_{wzz} = m'' > 0$; consequently, $\gamma_2 = C_{wzz} = m'' > 0$ and reserve uncertainty causes expected exploratory and development effort to decline more rapidly than the certainty-equivalent. Note also that $E_{qqq} = C_{www} = 0$ in equations (33) and (34) implies no effect of *price* uncertainty on the expected dynamics of decisions, as indicated by equations (29) and (31).

If we further assume that no additions to reserves are undertaken, then the way in which uncertainty about reserves affects extraction can be determined under slightly weaker hypotheses about extraction cost than equation (33). If there are no additions to reserves, then the expected dynamics of extraction are given by equations (29) and (30) with $q_z = 0$. In this case, it can be shown that $\gamma_1 > 0$ if $E(q, R) = qh(R) + \beta(q)$ with $E_{qRR} = h'' > 0$ and $E_{qqq} = \beta'' \geq 0$ (in other words, marginal cost is convex in the extraction rate). This specification includes equation (33) as a special case with $\beta(q) = (\beta/2)q^2$. If $\beta''' > 0$, then uncertainty about both price and reserves leads to more rapid expected depletion.

The above discussion also highlights sufficient conditions for the certainty equivalence approach to be applicable—that is, for uncertainty to have no effect on the expected dynamics of decisions. From equations (29) and (30) it follows that, in general, $\mathscr{E}_t\, dq/dt = d\bar{q}/dt$ if and only if $E_{qqq} = E_{qqR} = E_{qRR} = 0$. The first two conditions imply that the slope of marginal extraction cost is independent of both the extraction rate and reserves; the third condition implies that extraction cost is a linear function of the reserve stock. Taken together, these conditions imply that marginal cost is a linear function of q and R, which implies in turn that extraction cost satisfies the linear-quadratic form illustrated by equation (2). (With reference to equation (33), the conditions require that $h(R)$ be linear as in equation (2), where $h(R) = \eta - \gamma R$.) A similar argument shows that, in general, the certainty equivalence approach is applicable to exploratory effort and additions to reserves if and only if the cost function C has the linear-quadratic form of equation (17).

Our final topic concerns the response to uncertainty about reserves when the variance of reserves declines as depletion progresses, a form of "learning by doing." For brevity, we discuss in detail only the simple case of extraction with no accompanying additions to reserves; see Pindyck (1980) for a discussion of learning from exploration. Suppose that $\sigma_1 = \sigma_1(R)$ in equation (26), with $\sigma_1'(R) > 0$ so that uncertainty diminishes as reserves fall; assume also that $w = 0$.

As in Pindyck (1980, appendix B), it can be shown that

$$2E_{qq}[\mathscr{E}_t\, dq/dt - d\bar{q}/dt] = -\left[\gamma_1\sigma_1^2 + E_{qqq}q_p^2p^2\sigma_2^2\right] - 2\sigma_1\sigma_1'E_{qR} \quad (35)$$

where γ_1 is as in equation (30) with $q_z = 0$. Under the assumptions that $\sigma_1' > 0$ and $E_{qR} < 0$ (marginal cost rises as reserves fall), the "learning term" $-2\sigma_1\sigma_1'E_{qR} > 0$ in equation (35) acts to offset the tendency of convex costs to induce more rapid depletion relative to the certainty-equivalent; this latter influence is represented by the first term on the right-hand side of equation (35). Intuitively, the fact that depletion reduces uncertainty compensates (in present value terms) for the "higher average cost under uncertainty" effect described in the text; thus learning tends to moderate the output decline. This observation also leads to the conjecture that learning by doing about reserve additions also reduces the incentive for exploration and development, since the information value of exploration is smaller.

5
Aggregate Supply and Market Behavior

With this chapter, we shift our emphasis from decisions of individual firms to the determinants of resource supply and prices in a market context. The transition from the individual to the aggregate level of analysis is necessary because data on individual decisions generally are not available. Thus, empirical analysis of supply must be conducted using data that reflect both aggregation across individual participants and interactions among participants in various markets. In addition, empirical interest usually centers on the behavior of resource industries and markets rather than the decisions of individual participants.

An "aggregation problem" arises because the theory of resource supply characterizes the behavior of a representative firm working a representative deposit, while the data record the activities of many firms working many deposits. Aggregate data may not accurately represent the activities of a single firm; or, conversely, rules for individual decisions may be an incomplete description of market behavior. The gap between theory and data, the subject of this chapter, often compels researchers to supplement their conceptual models with *ad hoc* and sometimes questionable assumptions.

The aggregation problem, although common in the study of market behavior, is particularly serious in the study of resource supply. Resource deposits differ across both space and time in the size of reserves, degrees of depletion, geophysical conditions, extraction technology, and market conditions (including taxation and regulatory constraints). These underlying differences can lead to heterogeneities among individual decision rules in their form (for example, linear versus nonlinear), content (which variables influence decisions), and measures of the response of supply activities to common influences (such as prices or depletion effects). These heterogeneities in turn lead to differences between individual and aggregate behavior that typically defy expression as simple sums or averages.

A perspective on the logical and practical inconsistencies between individual decisions and market behavior can be gained by recalling the context of the theory developed in chapter 2. There we noted that individual decision rules could be derived as a solution to an intertemporal present-value-maximization problem. In deriving the solution, variables such as the resource price path, input costs, initial reserve stocks, geophysical conditions, and resource supply technologies were assumed to be fixed. The same approach cannot be applied to derive industry decision rules at the market level because these variables, particularly prices, are themselves determined by the actions of individual participants. Instead, the derivation of a market solution requires separate hypotheses about demand, supply, and how market interactions determine observed prices and quantities.

Conceptual descriptions of market interactions are deficient in economics generally, not just in the area of nonrenewable resources. With few exceptions, theories of markets characterize equilibrium outcomes that tie individual decisions to market clearing requirements, in contrast to descriptions of the processes by which markets determine prices and quantities. The few exceptions deal with special cases of price and quantity determinations in monopoly or oligopoly markets.[1] All of the competitive market models discussed in this chapter presume that an "invisible hand" is operating to determine equilibrium outcomes. The practical import of this assumption is an empirical matter—actual markets may closely approximate the actions of the invisible hand—but the narrow context should be kept in mind through subsequent discussions of market models.[2]

The need for a separate demand hypothesis to explain market behavior also introduces a number of practical problems in identifying the separate influences of supply and demand in market data. Again, the

1. See, for example, Zabel (1982), Miller (1982), and references cited therein.

2. Competitive market outcomes can be derived from models of social welfare maximization problems (Stiglitz, 1974a, 1974b; Toman, 1982c) but only under stringent assumptions including complete forward and insurance markets (Arrow, 1971, essays 4 and 5).

identification problem is common in empirical economics, but the usual estimation procedures may be especially deficient in dynamic resource models, where depletion and expectations play a crucial role in the decision process. This issue underlies the "rational expectations" approach to characterizing a market equilibrium and identifying supply responses, as discussed later in this chapter.

HETEROGENEITY AND AGGREGATION PROBLEMS

Available data on resource supply activities reflect substantial aggregation across firms, geographic regions, and grades of deposits as measured both by physical abundance and geophysical attributes that influence extraction and finding cost.[3] The question to be addressed is how useful models of individual behavior described in earlier chapters are in the analysis of aggregate supply data. This section addresses the question by examining the conditions for "perfect" aggregation. The reader will probably not be surprised to learn that the conditions fail in practice and that the answer to the question is ultimately an empirical matter. Nevertheless, it is useful to derive specific implications of the aggregation problem for evaluating and constructing empirical models, since a pragmatic balance between the desirable and the practical is required.

Starting with extraction behavior, one way to address the aggregation issue is to investigate what conditions are required for the aggregate cost of extraction to have the same content and properties as the individual cost function $E(q, R)$ found in chapter 2. Analyses by Solow and Wan (1976) and Hartwick (1978, 1982) indicate that two conditions are sufficient for an aggregate cost function to have the same content as the individual cost function described previously.

1. There is a range of deposits of different grades or qualities distinguished by different unit costs.
2. Unit cost for each deposit is invariant not only with respect to changes in the output rate but also with respect to declines in reserves.

Conditions (1) and (2) imply that a one-to-one correspondence exists between cumulative past extraction (and thus remaining reserves) and the grade of resource being mined at any particular time. They imply also

3. Chapter 6 describes some characteristics of the data available for analyzing the U.S. petroleum industry, which are perhaps more extensive than the data on any other resource industry.

that lower cost grades are mined first, so that aggregate unit cost rises as reserves fall, though unit cost is independent of the extraction rate.[4] The user cost of extraction may be interpreted as the present value of increases in unit costs during future periods as a consequence of currently depleting lower cost deposits. In the presence of a continuum of resource grades, unit cost will rise smoothly as total reserves decline. In the presence of a discrete number of resource grades, unit cost will rise (with a jump) at discrete moments in time with the transition from lower to higher cost deposits.[5]

Condition (2) is an implausible description of resource deposits, particularly petroleum, because extraction cost for an individual deposit may be expected to rise as extraction accelerates or as reserves decline. In this event, aggregate output at any moment will include output from deposits with different unit costs.[6] Consequently, both the marginal and the average cost schedules for aggregate extraction will be the horizontal sum of the corresponding schedules for individual deposits, as illustrated in figure 5-1.

Figure 5-1 indicates how aggregate cost schedules depend on the position and slope of the individual cost schedules, as well as on aggregate extraction rate and reserve volume. The position and slope of the individual schedules will depend in turn on the initial size of deposits, their particular geophysical characteristics, and their "vintage"—that is, the degree of past depletion. In particular, the vintage determines whether cost is sufficiently low that capacity operation is optimal, in contrast to extraction along a decline curve in later stages of operation. It follows that aggregate cost, and hence aggregate supply behavior, will depend on the distribution of deposits according to these various attributes, in addition to the magnitude of aggregate reserves and the level of current and future prices. Moreover, these distributional influences will change over time, as older deposits are abandoned and new reserves are developed, in ways that are reflected inadequately (if at all) in specifications of the rise in costs for a single deposit as reserves decline.

The same problems arise in using models of individual exploration and development decisions to describe industry behavior. Of particular

4. Solow and Wan (1976, footnote 3). Note that aggregate cost has the same general properties as the constant marginal cost model discussed in chapter 2. Thus, condition 2 can be replaced by an alternative condition 2′: unit cost for each deposit increases as reserves decline, but extractive capacity is independent of reserves, and reserves of each grade are produced at maximum capacity to the point of complete physical exhaustion. Consequently, reserves of each type must be exhausted prior to the point at which extraction follows the "decline curve" q^* in figure 2-3a.

5. Hartwick (1982, pp. 285–287).

6. Hartwick (1978, pp. 215–216).

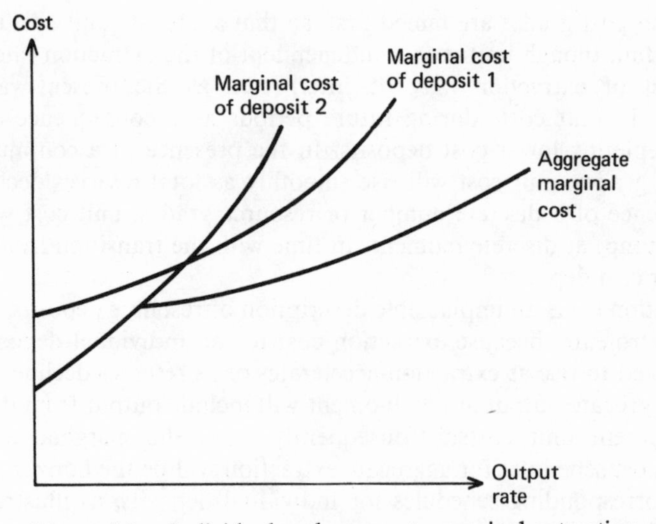

Figure 5-1. Individual and aggregate marginal extraction cost.

importance is the specification of depletion effects at the development and exploration stages. For the individual firm described in chapter 2, depletion effects depend on cumulative discoveries or reserve additions by the firm alone. When the firm's proportion of total activity in a region is small, the firm may view its decisions as having essentially no effect on exploration and development costs. The actions of other firms cannot be ignored, however, and represent an exogenous influence on costs that the firm must forecast to determine an optimal plan. These interactions may also give rise to common property externalities as described in chapter 3, which alter individual behavior in the presence of other firms and cause aggregate behavior to deviate from the sum of individual actions taken independently of the rest of the industry. Models of individual behavior also fail to take into account interactions among firms that determine the value of reserves, the probability of new discoveries, and other measures of industry performance discussed below.

Another approach to the heterogeneity problem concerns the aggregation of decision rules, in contrast to the aggregation of costs that underlie the decision rules. Consider the extraction decision rule derived in chapter 2, equation (13). That decision rule can be written as

$$q_{it} = f_i(p_t, R_{it}, x_t) \tag{1}$$

where p_t is the current price of extracted output, R_{it} is producer i's reserve stock, and x_t is a vector of other influences common to all firms.

These common influences may include measures of technical progress, factor prices, and past resource prices and costs that influence expectations, as described in chapter 4. The elements of x_t may be nonstochastic or random. A corresponding hypothesis about aggregate extraction based on the theory of individual behavior is represented by the equation

$$q_t = f(p_t, R_t, x_t) \tag{2}$$

where $q_t = \sum_i q_{it}$ is aggregate output, $R_t = \Sigma R_{it}$ is aggregate reserves, and f is a function that has the same general form as the individual supply functions f_i in equation (1). The aggregate expression in equation (2) is motivated by the use of industry-level data in a model of individual firm behavior. Alternatively, averages of aggregate variables could be inserted in equation (1) to describe the behavior of the representative firm.

Two conditions are needed to avoid aggregation problems in this setting. The first is that the individual supply functions are linear, so that

$$q_{it} = A_i + B_i p_t + C_i R_{it} + D_i x_t \tag{3}$$

where A_i, B_i, and C_i are fixed parameters and D_i is a vector of fixed parameters that is conformable with x_t. If the f_i are nonlinear, then the form of the aggregate function f will be the same as the forms of the f_i only by coincidence. Using equation (3), aggregate output can be represented by

$$q_t = A + Bp_t + \sum C_i R_{it} + Dx_t \tag{4}$$

where $A = \Sigma A_i$, $B = \Sigma B_i$, and $D = \Sigma D_i$ (for each term in the parameter vectors D_i and D).

The second condition follows from equation (4): the reserves coefficient C_i must be the same for all producers, so that

$$q_t = A + Bp_t + CR_t + Dx_t \tag{5}$$

where $C = C_i$ for all i. If the C_i are not identical, then aggregate behavior will depend on the distribution of individual producers' reserves, rather than depending only on aggregate reserves.

The disparity between equations (4) and (5) is an example of the classic aggregation problem.[7] The problem is compounded by the presence of other firm-specific influences on supply besides reserves.[8] One example indicated in chapter 2 is the cumulative volume of a firm's additions to reserves, which reflects interactions among its extraction, development, and exploration decisions. Examples pertinent to the petroleum industry are provided in chapter 6. To avoid aggregation problems, the coefficients of firm-specific influences also must be identical across individual firms.

7. See, for example, Theil (1971, chapter 11).

8. In that case, the term $D_i x_t$ in equation (3) is replaced by $Dx_t + F_i z_{it}$, where z_{it} is a vector of other firm-specific influences.

A sense of the importance of the aggregation problem is gained by noting that equality of the reserves coefficients C_i across individual decision rules requires very stringent assumptions on the underlying cost parameters. To see this, suppose that individual costs satisfy the linear-quadratic specification

$$E^i(q_i, R_i) = (a_i - c_i R_i) q_i - (b_i/2) q_i^2 \tag{6}$$

discussed in chapter 4, where $a_i > 0$, $b_i > 0$, and $c_i > 0$ are fixed parameters. For simplicity, assume that the individual discount rates r_i and the price p of output do not change over time. Under these assumptions, it can be shown that optimal output for firm i satisfies[9]

$$q_{it} = A_i + B_i p + C_i R_{it} \tag{7}$$

where the coefficients A_i, B_i, and C_i satisfy

$$\begin{aligned} A_i &= K_i a_i / c_i \\ B_i &= K_i / c_i \\ C_i &= K_i \\ K_i &= \tfrac{1}{2}\left(r_i - \sqrt{r_i^2 + 4 r_i c_i / b_i}\right) \end{aligned} \tag{8}$$

It follows from these equations that the C_i will be identical only if the discount rate and the underlying cost parameters c_i and b_i are identical for all firms. The latter condition implies that all individual marginal cost schedules have the same slope and that equal changes in reserves have the same effect on marginal cost for each individual deposit. These are very strong assumptions. Note also that these assumptions imply identical price response coefficients B_i for all firms. Consequently, the individual supply functions given by equation (7) are constrained to differ only in their intercept coefficients A_i, through differences in the individual cost parameters a_i. These results are important in the discussion of empirical models of supply behavior in chapter 6.

As serious as these observations indicate the heterogeneity problem may be, they nevertheless understate the problem because individual decision rules can differ in form as well as in content. In particular, individual decision rules will vary according to whether capacity constraints are binding and according to incentives for temporarily deferring extraction or additions to reserves (that is, binding nonnegativity constraints).[10] Moreover, individual decision rules change as depleted

9. See Toman (1982b) for the derivation.

10. Capacity constraints introduce a difference in content as well as form, with capacity-constrained decisions being influenced by determinants of capacity such as durable equipment investment.

deposits are abandoned and as new deposits are brought into operation. Aggregate supply will depend on the composition of these factors as well as on the determinants of individual decisions; yet, compositional factors are ignored in descriptions of aggregate behavior based solely on decision rules of the representative individual.

MARKET INTERACTIONS

Interactions among demand and supply decisions are the second principal source of divergence between individual and aggregate supply behavior. The divergence occurs because important determinants of supply behavior, such as output prices and input costs, are exogenous to decisions at the individual level, yet are determined by the aggregate of decisions at the market level. Our discussion of this issue begins with an overview of the dynamics of resource demand as a prelude to the interactions between demand supply in final output (extracted resource) markets and in intermediate markets for inputs, including new discoveries and reserves. The section concludes with brief remarks on inventories and on markets for mineral rights.

Dynamics of Demand

Our objective is not to provide an exhaustive treatment of nonrenewable resource demand in general, or even energy demand in particular, but to highlight the inherently intertemporal nature of resource demand behavior.[11] A distinguishing feature of the demand for nonrenewable resources is that it is typically a derived demand for the services these resources provide as inputs to the production or consumption of other commodities. Energy resources, for example, are required to provide space heat, transportation fuel, and so forth. The importance of this derived-demand characteristic stems from the fact that resource inputs are used in combination with durable capital stock inputs (such as automobiles, household appliances, boilers, generators, and the like). Resource demand thus depends on the rate at which existing capital stocks are used (a short-run phenomenon) and on investment in new capital stocks (a long-run phenomenon). Because capital stock investment decisions are inherently dynamic, and because capital and resources are used together, decisions about resource demand inherit a dynamic structure from investment decisions.

11. See Bohi (1981) for a critical review of energy demand analyses.

This point can be illustrated with a simple dynamic model of an individual resource-using firm.[12] Suppose that production of the firm's output can be represented by a production function $G(K, q, L)$, which depends on its capital stock K, resource use q, and labor (or other variable factor) input L. Let I denote the rate of capital investment, p the price of the resource input, and w the price of the labor input. Assume that the cost of investment can be represented by a cost function $mI + A(I)$, which combines the cost of the investment good mI and the "adjustment costs" $A(I)$ incurred by the firm in changing the size of its capital stock. For simplicity, assume that the price of the firm's output is constant over time and that input prices are normalized so that the price of final output is equal to unity (thus p, w, and m represent input prices relative to the price of output, and $A(I)$ the relative cost of adjustment). Also assume for simplicity that the capital stock does not depreciate over time.

With these assumptions, the present value of the firm's net revenues is given by

$$V = \sum_{t=0}^{T} d^t [G(K_t, q_t, L_t) - p_t q_t - w_t L_t - m_t I_t - A(I_t)] \qquad (9)$$

where d is the discount factor and T is the firm's operating horizon. Factor demands are derived by maximizing this expression subject to the capital stock equation $K_t = K_{t-1} + I_t$, where K_{t-1} is the stock of capital on hand at the start of period t.[13] Ignoring corner conditions, and letting G_K, G_q, and G_L denote marginal products (that is, derivatives of the production function), optimal factor demands will satisfy

$$G_q(K_{t-1} + I_t, q_t, L_t) = p_t \qquad (10)$$

$$G_L(K_{t-1} + I_t, q_t, L_t) = w_t \qquad (11)$$

$$G_K(K_{t-1} + I_t, q_t, L_t) + \sum_{s=t+1}^{T} d^{s-t} G_K(K_s, q_s, L_s) = m_t + A'(I_t) \qquad (12)$$

Note that G_q and G_L are the values of marginal product of q and L, respectively, since the output price has been normalized to one. Thus, equations (10) and (11) are the standard efficiency conditions for profit maximization: for each factor, they equate factor prices with the value of marginal product. Note also that the marginal products depend on the capital stock inherited from the previous period, K_{t-1}, and on the

12. This model is a discrete-time analogue of Treadway's (1970) model, with a nonrenewable resource included as a factor of production. A similar (continuous-time) model also is sketched in Bruno and Sachs (1982).

13. Note that we assume no lag between the investment decision and additions to capital.

investment rate I_t. Equation (12) is the dynamic efficiency condition for capital investment. The first term on the left-hand side is the value of the marginal product of current investment. The second term is the shadow value of new investment and is equal to the present value (at time t) of future marginal products of capital. This term, which is analogous to the user cost expressions derived in chapter 2, appears in equation (12) because current investment also yields services in future periods. Therefore, equation (12) states that along an optimal investment path the full marginal value of investment, including the value of future services, is equal to marginal investment cost in each period.

The backward recursion approach described in chapter 2 can be used to derive dynamic factor demand equations from the efficiency conditions stated in equations (10) through (12). Without going into details, we note first that equations (10) and (11) imply that resource and labor demands depend on current input prices (p_t, w_t), existing capital K_{t-1}, and capital additions I_t. Equation (12) implies that I_t depends on m_t, K_{t-1}, q_t, L_t, and on future magnitudes of these variables through the shadow value term $\Sigma d^{s-t} G_K$. Since future factor demands ultimately depend on future input prices, it follows that I_t also depends on current and future prices of all inputs. The same is true of q_t and L_t in view of equations (10) and (11). Therefore, equations (10) through (12) imply that factor demands satisfy the following general functional relationships:

$$q_t = q(p_t, p_{t+1}, \ldots; w_t, w_{t+1}, \ldots; m_t, m_{t+1}, \ldots; K_{t-1}) \qquad (13)$$

$$L_t = L(p_t, p_{t+1}, \ldots; w_t, w_{t+1}, \ldots; m_t, m_{t+1}, \ldots; K_{t-1}) \qquad (14)$$

$$I_t = I(p_t, p_{t+1}, \ldots; w_t, w_{t+1}, \ldots; m_t, m_{t+1}, \ldots; K_{t-1}) \qquad (15)$$

The dynamics of resource and other factor demands summarized by equations (13) through (15) are similar to the dynamics of supply examined in chapter 2. In particular, these equations indicate that interactions among demand decisions are analogous to the interactions among supply decisions at different stages. Resource demands depend on relative factor prices and on expectations of relative factor prices. Thus, expectations of changes in the price of q relative to L or I will change current investment decisions, which in turn will change current demands for L and q. With more than one type of capital good, expectations of relative factor price changes will affect factor demands through changes in the composition of the capital stock. Changes in factor prices relative to the price of output also will alter factor demands through changes in the scale of production. Finally, the direction (as well as the magnitude) of the response of resource demands to price changes will depend on the structure of technology, specifically on whether q is a substitute for, or a complement to, L and K.

The dynamic analysis of demand can be extended to substitution possibilities among resources. Suppose that factor L is interpreted not as a labor input but as another nonrenewable resource that is a substitute for q. It follows from equations (13) and (14) that the resources are joint products in demand, with demand for each being a function of their relative prices. Possibilities for substitution may exist only in the longer run with changes in capital, as in utilities' decisions to operate coal-fired rather than oil-fired generating equipment. Alternatively, substitution possibilities may exist even in the very short run, as in industrial demand for dual-fired boilers that can be switched rapidly between residual fuel oil and natural gas. These possibilities will be reflected in the dependence of resource demands on the composition as well as on the size of capital stocks.

Another extension of the analysis concerns the availability of "backstops" that can be substituted for nonrenewable resources and are not themselves subject to depletion effects over time.[14] Suppose that factor L is interpreted as such a backstop input, rather than as labor. Then expectations of falling backstop prices (a decline in w_t over time), perhaps due to technical innovation in backstop production, will alter current resource demand through the effect on capital stock decisions.

Although this discussion of resource demand has been couched in terms of a firm's behavior, the same reasoning applies to household demand. For household demand, the production function is replaced by a utility function, and the objective is to maximize discounted total utility subject to budget constraints reflecting household income, prices of consumption and household investment goods, and costs (pecuniary or psychic) that result from changes in household capital or expenditure patterns. The derived demand for resources will depend on current and expected future prices of the resource relative to other goods (including other resources), income, interest rates, and household capital stocks. Capital in this case includes both consumer durables and investments in greater efficiency, such as weatherstripping, which alter relative resource demand.[15]

Interactions Between Supply and Demand for Extracted Output

Interactions between market demand and supply decisions for extracted resources occur in two dimensions. First, changes in demand feed back on supply behavior through their effect on the current market price and vice versa. The second dimension involves dynamic interactions between decisions through changes in expectations and changes in resource stocks

14. See, for example, Dasgupta and Heal (1979, chapter 7).

15. As discussed below, resource demand and supply also depend on institutional factors, such as taxes and subsidies, and on expectations of changes in these factors.

and capital stocks. Because demand depends on expectations of future prices, changes in expectations will influence current demand decisions and thus feed back on supply. Similarly, changes in producers' price expectations will feed back on demand. Changes in resource stocks and capital stocks have analogous dynamic effects.

These interactions can be illustrated by the simple market model specified in the following four equations:

$$q_t = q(p_t, p_{t+1}^e, p_{t+2}^e, \ldots; m_t, m_{t+1}^e, m_{t+2}^e, \ldots; K_{t-1}) \tag{16}$$

$$I_t = I(p_t, p_{t+1}^e, p_{t+2}^e, \ldots; m_t, m_{t+1}^e, m_{t+2}^e, \ldots; K_{t-1}) \tag{17}$$

$$q_t = f(p_t, p_{t+1}^e, p_{t+2}^e, \ldots; R_t, z_t) \tag{18}$$

$$y_t = g(p_t, p_{t+1}^e, p_{t+2}^e, \ldots; R_t, z_t) \tag{19}$$

where y represents new additions to reserves R, and z represents cumulative past additions, as in chapter 2. Equations (16) and (17) are simplified versions of the more general input demand functions (13) and (15), while equations (18) and (19) correspond to the supply functions given by equations (32) and (33) in chapter 2 for an integrated resource producer. In all four equations, decisions depend on expectations of future resource prices, $p_{t+1}^e, p_{t+2}^e, \ldots$, on the current price p_t, and on the stock variables (K_{t-1}, R_t, z_t). Resource demand and capital investment decisions also depend on the expected path of prices for new capital goods $(m_t, m_{t+1}^e, \ldots)$. For simplicity, we assume that these prices are exogenous, though in practice, interactions between supply and demand decisions for capital stocks should be addressed.

Under the hypothesis of market equilibrium, equations (16) and (18) jointly determine the current price and quantity of extracted output for given magnitudes of the stock variables and price expectations. Holding expectations fixed, the equilibrium will change over time with adjustments to the capital stock and additions to the stock of reserves, as given by equations (17) and (19). If price expectations are hypothesized to depend only on current and past prices, then the system can be rewritten as

$$q_t = q(p_t, p_{t-1}, p_{t-2}, \ldots; m_t, m_{t-1}, m_{t-2}, \ldots; K_{t-1}) \tag{20}$$

$$I_t = I(p_t, p_{t-1}, p_{t-2}, \ldots; m_t, m_{t-1}, m_{t-2}, \ldots; K_{t-1}) \tag{21}$$

$$q_t = f(p_t, p_{t-1}, p_{t-2}, \ldots; R_t, z_t) \tag{22}$$

$$y_t = g(p_t, p_{t-1}, p_{t-2}, \ldots; R_t, z_t) \tag{23}$$

with lagged values of prices replacing expected future values.

By choosing specific functional forms, a dynamic simultaneous-equations regression model can be derived from equations (20) through (23). Since market data reflect interactions between demand and supply decisions, a joint estimation procedure is needed to sort out these interactions and distinguish supply responses.[16] Identification is achieved under the hypothesis that price expectations depend only on past prices, as the demand-specific capital stock variable K_{t-1} and the supply-specific resource stock variables R_t and z_t are available to distinguish changes in demand from changes in supply. In other words, the system given by equations (20) through (23) contains so-called exclusion restrictions that can be used to identify the equations in the model.[17]

However, the model is underidentified according to the criterion of exclusion restrictions if the demand-side capital stock variable influences supply through its effect on price expectations or if consumer price expectations depend on resource reserves. The first influence follows if resource producers forecast future resource prices recognizing the connection between capital stocks and demand, while the second relationship holds if resource consumers forecast future resource prices recognizing the connection between reserves and supply. If these influences are present, then the demand and supply functions depend on all the variables in the system and there are no arguments specific to each equation to distinguish one from the other. It follows that an alternative hypothesis to exclusion restrictions is needed to achieve identification. One such hypothesis is suggested by the concept of rational expectations market equilibrium discussed below.

Illustrations of Market Dynamics

Few examples exist in the resource supply literature to illustrate the importance of demand on market supply behavior. In the empirical literature on petroleum supply discussed in chapter 6, for example, few studies attempt to integrate hypotheses about demand with the supply models. The theoretical literature is only slightly less deficient in this respect. When demand is brought into the picture, most theoretical supply models assume a stationary demand function of the form $q = q(p)$, where p is the current price of the resource, and ignore the sensitivity of

16. Note that the system contains lagged endogenous prices and quantity variables through the dependence of stocks on past decisions. Several estimation procedures are available to address this lagged dependence, according to whether the error terms are serially correlated or correlated with the explanatory variables. See, for example, Theil (1971, chapter 10) and Granger and Newbold (1977, chapters 6 and 7).

17. See Fisher (1966) for a further discussion of exclusion restrictions and the identification problem.

demand to future price expectations, capital stocks, and relative prices of other goods.[18]

Despite this limitation, however, existing conceptual models can be used to shed some light on the dynamics of market behavior. Pindyck (1978c) investigates the dynamics of resource supply, price, and exploration under the assumption that price follows a market equilibrium trajectory, but where resource demand depends only on the current price.[19] In Pindyck's model, the marginal cost of reserve additions is an increasing function of reserve additions, while marginal extraction cost varies inversely with the level of reserves and is independent of the rate of extraction.

Pindyck's findings are summarized in figure 5-2.[20] Panel (a) shows the paths of reserves and "exploratory effort" when initial reserves are large. Large initial reserves imply a low extraction cost and a high initial rate of extraction, while exploratory effort is deferred until later periods. Reserves decline over time, though the rate of decline is tempered as exploration rises to a peak before subsequently declining. The extraction rate declines from its initial high level, and the price of extracted resource rises. A reduction in extraction costs for any level of reserves shifts the paths of both reserves and exploratory effort to the right, as indicated by the dashed curves in the diagram. One interpretation of these shifts is that they represent the effects of technical progress. Alternatively, the shifts can be interpreted in terms of changes in the mix of geophysical characteristics of the deposits being operated.

Panel (b) of figure 5-2 shows reserves and decision paths when initial reserves are small. In this case, the cost of extraction and the shadow price of new reserves initially are high, so the extraction rate is low, exploratory effort is high, and reserves grow. As reserves grow the rate of extraction increases (since extraction cost falls), and the shadow price of new reserves declines and causes exploratory effort to fall off. Reserves and extraction ultimately hit a peak and subsequently decline, while the price follows a U-shaped path—declining in the initial phase as production rises, then increasing as production falls. As shown by the broken lines in figure 5-2b, lower extraction cost for any level of reserves reduces exploratory effort—since smaller extraction cost implies a smaller shadow

18. Two exceptions are Becker (1980), who considers a general equilibrium model but ignores depletion effects inherent in resource supply, and Toman (1982c), who considers a partial equilibrium model with depletion effects. However, these studies are concerned primarily with existence of market equilibrium rather than qualitative findings. Studies that yield qualitative conclusions, such as Stiglitz (1974a, 1974b), tend to neglect depletion effects.

19. The demand hypothesis is explicit in Pindyck's (1980) extension to incorporate uncertainty.

20. Adapted from Pindyck (1978c, pp. 850–851).

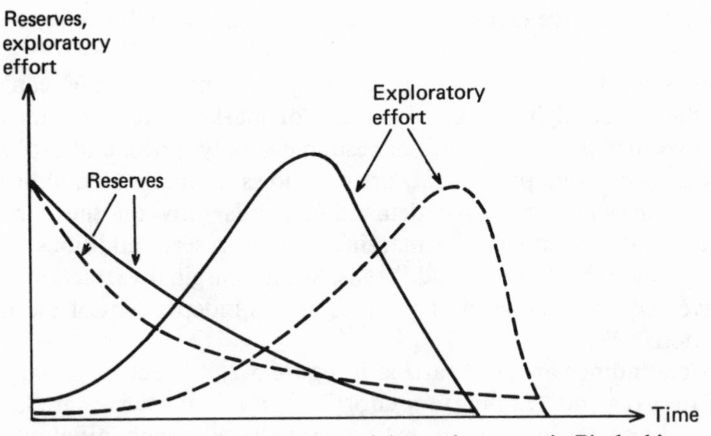

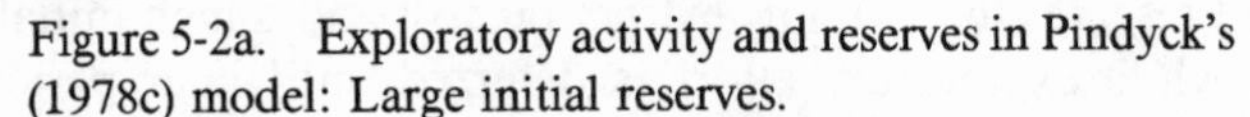
Figure 5-2a. Exploratory activity and reserves in Pindyck's (1978c) model: Large initial reserves.

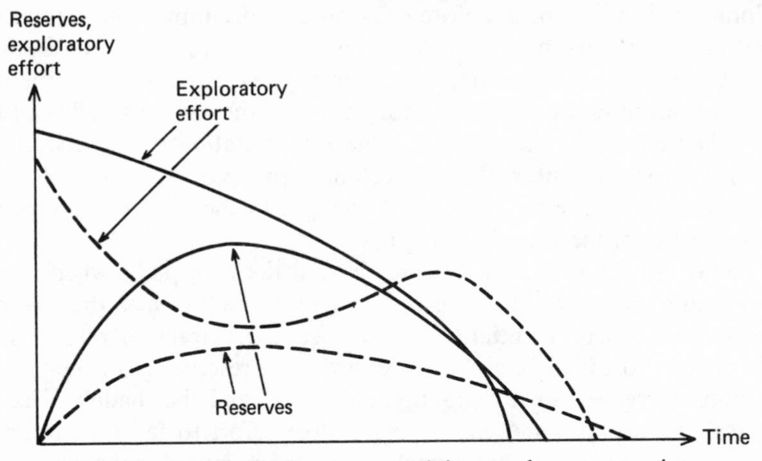

Figure 5-2b. Exploratory activity and reserves in Pindyck's (1978c) model: Small initial reserves.

price of new reserves—and may cause a departure from a path of monotonic decline. Comparison of panels (a) and (b) illustrates how market outcomes vary with different levels of reserves.

Broadly similar conclusions are reached by Liu and Sutinen (1982), as summarized in figure 5-3 for the case of low initial reserves and no past discoveries.[21] In contrast to Pindyck's model, the marginal cost of additions to reserves is assumed to be constant (instead of an increasing function of reserve additions) while marginal extraction cost increases

21. Adapted from Liu and Sutinen (1982, p. 156).

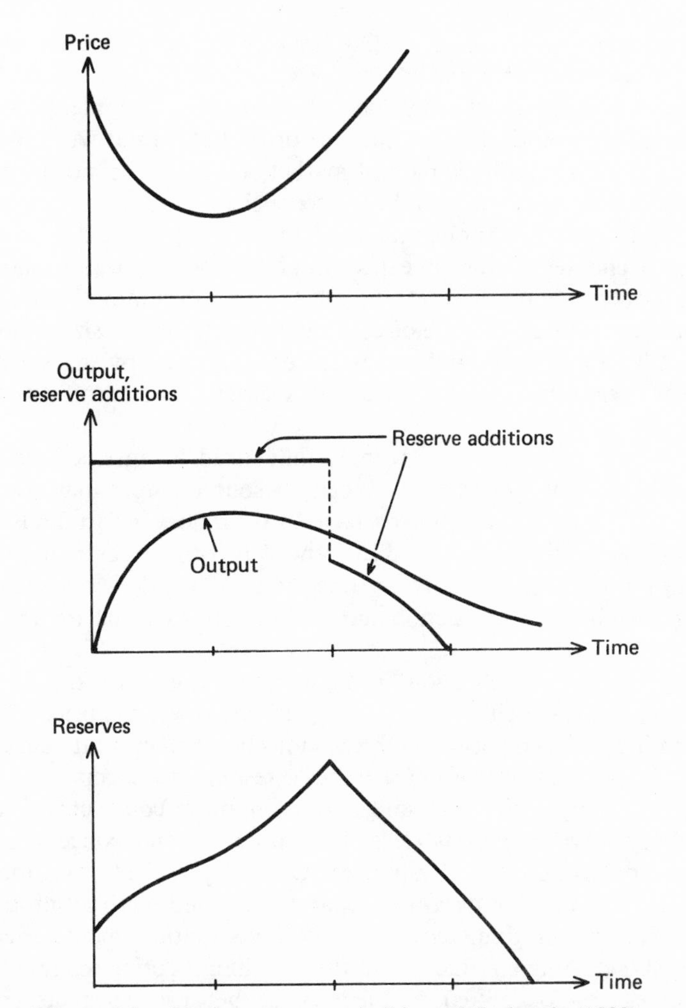

Figure 5-3. Market outcomes in Liu and Sutinen's (1982) model.

with the rate of output (instead of being independent of the output rate). As a consequence of these differences in assumptions, the path of reserve additions in figure 5-3 follows an initial phase at a constant capacity rate, and then abruptly jumps down along a declining path.[22] Extraction rises initially but is outpaced by reserve additions, so the stock of reserves increases. Ultimately, however, extraction outstrips reserve additions and the stock of reserves declines. The output path in figure 5-3 implies a

22. The path of reserve additions is similar to the extraction path in figure 2-3, where analogous assumptions about cost are made.

U-shaped path of prices. If initial reserves are large, output declines monotonically and price rises.[23]

Three important conclusions emerge from these examples. First, it is difficult to determine whether the price or rate of output will follow any specific path. The predicted path of market outcomes depends on costs of production and exploration and on reserve sizes, among other influences. Thus, predictions of exponential price growth based on the "*r* percent rule" (see chapter 2), or of exponential decline in output, should be viewed with suspicion. Second, the market equilibrium path will be in a constant state of flux because of continual changes in the size and cost of reserves. These dynamic influences are not captured by market models that characterize market equilibrium as a series of static market-clearing conditions.

Third, the market path is strongly influenced by interactions among stages in the supply process. Specifically, resource output may rise or fall over time depending on the paths of exploratory effort and reserve additions, while these paths are influenced in turn by extraction incentives. Studies that focus only on one stage of supply or assume rigid relationships between extraction and exploration decisions are unlikely to capture these interactions.

Before closing this section, it is worth noting that the foregoing discussion focuses exclusively on spot markets, where the price of output is determined in each period by interactions between current demand and supply. In practice, many nonrenewable resources—specifically natural gas, coal, and uranium—are sold under long-term contract.[24] The presence of long-term contracts raises both practical and conceptual problems in analyzing supply responses. On the practical side, there are difficulties in obtaining data that separate spot and contract production, record contract duration, and distinguish between the average price (over all new and existing contracts) and the incremental price for spot or new contract sales. Conceptual problems arise because contracts create a distinction between the economic incentives pertaining to new and existing contracts. In the limiting case of rigid contracts, extraction decisions linked to existing contracts will be unresponsive to changes in incentives on new contracts, while exploration and development decisions will be

23. If past discoveries are sufficiently large, then the marginal cost of additional discoveries will exceed their value. In this case, reserve additions are zero for all t and extraction declines monotonically in figure 5-3.

24. Long-term contracting is particularly significant in natural gas markets because of interactions between contract provisions and price controls. Off-system gas sales between pipelines are somewhat analogous to spot transactions, as are sales under short-term contracts to interruptible industrial users with dual-fired capability.

affected both by the change in incentives and by the extent to which contract rigidities blunt incentives.

Supply behavior can be influenced by contract terms in ways that are not adequately captured by a spot market model. For example, contracts and spot transactions generally differ in the distribution of risks between buyer and seller, though the relative risk burden depends on specific contract provisions.[25] Contracts also frequently involve middlemen acting as brokers and transporters—pipelines, in the case of natural gas—whose incentives and responses must be assessed to complete the picture of market behavior. Finally, contract terms are related to the incentives to acquire new reserves or hold inventories. These aspects of supply are addressed in the next section.

Supply and Demand Interactions in Intermediate Markets

In addition to the factors involved directly in the market for extracted output, observed behavior also reflects demand and supply interactions in intermediate markets for new reserves and for other factors of production involved in the supply process. Constraints imposed on resource supply by the availability of highly specialized inputs, such as rotary drilling rigs, can be especially significant.[26] When the constraints are binding, changes in resource supply imply an appreciable change in aggregate demand for inputs and their prices, thus invalidating the assumption of an exogenous input price.

Interactions among decisions to acquire and sell new discoveries and developed reserves also are important, yet information on these transactions is scanty and unreliable. Information on the size and value of reserves is highly proprietary and not widely reported in the same way as data on extracted output. Nominal similarities among deposits also may be overshadowed by their specific characteristics, creating serious problems in measuring prices and quantities. Moreover, reserve transactions often take indirect forms, such as horizontal merger or the purchase or sale of a working interest in a resource deposit. In addition to the lack of reported information, firms involved in transactions face substantial uncertainty in gauging the size and worth of reserves, and the uncertainty may be asymmetric between buyer and seller. The party with better

25. In natural gas markets, for example, "take-or-pay" provisions require buyers to pay for a certain percentage of contracted gas whether or not they actually use it, and "market-out" provisions allow buyers to terminate contracts unilaterally. Take-or-pay provisions shift the risk of demand fluctuations to buyers; market-out provisions shift this risk to sellers. See Broadman and Montgomery (1983) for a discussion of these issues and a general review of analytical problems posed by natural gas contracts and regulation.

26. Specialized labor skills, such as those possessed by petroleum engineers and geologists, also are important in this connection.

information may be able to exploit an advantage, creating a form of market failure that raises questions about whether a competitive model is an appropriate characterization of the market for reserves and discoveries, even if there are many buyers and sellers.[27]

With these issues in mind, consider the incentives for transactions among nonintegrated firms in a competitive market for new reserves. By analogy with the integrated-firm model developed in chapter 2, let $E(q, R)$ represent the cost of extraction to a nonintegrated producer and let $C(y, z)$ represent the cost of finding new discoveries to a nonintegrated exploration firm. Assume further that the nonintegrated extraction firm incurs a cost of $D(y)$ for developing new discoveries that are purchased from the exploration firm at a price of p^R per unit. Then using an argument similar to that applied in chapter 2, the following marginal profit rules can be derived for the two firms:

$$p_t - E_q(q_t, R_t) = p_t^R + D'(y_t) \qquad \text{(extraction firm)} \qquad (24)$$

$$p_t^R + D'(y_t) = \sum_{s=t+1}^{T} d^{s-t}[-E_R(q_s, R_s)] \qquad \text{(market clearing)} \qquad (25)$$

$$p_t^R = C_y(y_t, z_t) + \sum_{s=t+1}^{T} d^{s-t}C_z(y_s, z_s) \qquad \text{(exploration firm)} \qquad (26)$$

Equation (24) states that optimal behavior for an extraction firm is to balance the marginal profit of production against the marginal replacement cost of reserves, $p^R + D'(y)$. Equation (26), which is analogous to equation (26) in chapter 2, states that optimal behavior for an exploration firm is to equate the price of new reserves to their full marginal cost, including user cost $\Sigma d^{s-t}C_z$. Equation (25) is the market clearing condition: it states that if buyers and sellers correctly value new reserves, then the amount an extraction firm is willing to pay for new reserves equals their shadow value $\Sigma d^{s-t}[-E_R]$, less their marginal development cost $D'(y)$.[28] Note that combining equations (24) and (25) yields equation (3) in chapter 2, the marginal profit rule for an integrated extraction firm.

Equations (24) and (25) can be solved, at least in principle, to yield a supply function for extracted output and a demand function for new reserves, and equation (26) yields a supply function for new reserves. These behavioral equations will depend on current reserves and past discoveries (R_t, z_t), and on current and future prices of extracted output

27. One reason that firms elect to acquire new reserves through horizontal merger might be that this form of transaction provides a way to also obtain the seller's information about the reserves and thus avoid problems of "adverse selection."

28. See Pindyck (1982, pp. 293–294) for a similar argument.

and new reserves. Thus, the reserves market model exhibits the same dynamic interactions as the final output market model discussed above.

However, the lack of information on market transactions for reserves precludes direct implementation of the model. As noted in chapter 6 in connection with petroleum, data on reserves are spotty and are not defined consistently with the information requirements of the theory. The absence of reported data on the price of new reserves is even more serious. One might be tempted to estimate prices with equation (25) using measures of marginal development cost D' and depletion effects E_R.[29] Aside from the practical difficulties encountered in obtaining cost data, there is another drawback to this approach: using equation (25) to construct a time series for p^R removes from the model the independent equation required to identify the demand for reserves. Using estimated prices means that interactions between demand and supply for reserves cannot be distinguished because the constructed price data will combine both influences. Another problem is that restrictive assumptions about the interactions between extraction and reserve additions are required to use equation (25) in estimating prices of reserves.

An alternative way to analyze the market for reserves is to combine equations (24) through (26) to eliminate the unobservable price variable p^R and proceed as if all producers were fully integrated into exploration and development. However, this approach assumes market outcomes are identical to the decisions of competitive integrated firms. Hence, market interactions are ignored, as are influences that are specific to demand or supply decisions, such as information asymmetries.

Empirical analysis of the market for reserves involves several additional considerations, including mergers, joint interests, inventories, and mineral leasing procedures. Each of these considerations introduces another level of complexity in modeling the market, and a potential deviation from the market model described in equations (21) through (26). Concerning mergers and joint interests, Tourinho (1979) sketches a simple model of an "options approach," in which reserves represent a capital asset that provides an option on future revenues from extraction. The value of the option is inferred from hypotheses about the operation of capital markets. Although the scope of Tourinho's analysis is limited (specifically, depletion effects and supply dynamics in valuing reserves are ignored), the methodology seems promising.

The possibility of holding inventories of extracted output or developed reserves also introduces complications concerning risk preferences of suppliers. These inventories may be held to meet unexpected or seasonal peaks in demand, to provide a hedge against market fluctuations that

29. See, for example, Pindyck (1982, p. 297) and Epple (1983) for a discussion of this approach.

economize on the high front-end capital costs associated with new capacity, or to provide a means of spreading risks. Consequently, changes in transaction costs, risk preferences, and the degree of market instability will influence both demand and supply of new discoveries and existing reserves.[30]

The market for mineral rights also is important in countries like the United States where there is private ownership of resource-bearing land. Many studies treat resource-bearing land as fixed in supply, though of varying quality, and apply the concept of Ricardian rent to determine the quantity of land used and the value of inframarginal sites.[31] The approach is static and hence ignores the intertemporal considerations involved in demand and supply decisions for mineral rights. In particular, owners of resource-bearing land seeking to maximize the present value of net returns must consider the timing as well as volume of sale or lease decisions, and the opportunity cost of opening their land to resource exploration.[32] The current price of resource-bearing land will depend on current and future prices of the resource, and changes in the price of land will alter the incentives for holding land for alternative uses or for future resource exploration. The possibility of a price response is contrary to the view that the supply of resource-bearing land is fixed, and this possibility can make a difference in predictions about the responsiveness of new discoveries and the rate of extraction.[33] Predictions also will be sensitive to the proportion of publicly owned land being used and the processes by which mineral rights for these sites are allocated.[34]

Finally, leasing terms also feed back on behavior at the other stages of the supply process. Specifically, "diligence clauses" that require producers to exploit their holdings by a certain date or to surrender the lease will induce different exploration and production responses than firms would choose in the absence of contractual constraints, particularly in stagnant markets in which prices of outputs are falling. These responses

30. To cite one example, foreign expropriations of oil reserves in the early 1970s significantly increased the value of domestic U.S. reserves and stimulated domestic exploration, even though domestic price controls tempered incentives.

31. See McDonald (1979) for an extensive discussion of this approach.

32. Opportunity cost in this context consists of forgone current revenues from other uses, such as agriculture or residential development, and a user cost that reflects forgone future revenues from current (versus future) sale or lease.

33. Epple (1975) demonstrates how a nonzero elasticity of supply of oil-bearing land and a nonzero elasticity of substitution between oil-bearing land and other factors of production will alter the relationship between the supply of oil and rents to oil producers and landowners. Thus, estimates of supply responses based on rental income, such as Mancke (1970), are grossly oversimplified.

34. See, for example, Gilley and Karels (1981) and references therein for discussions of these issues from a theoretical and empirical perspective, particularly with respect to petroleum.

in turn will have an influence on both output prices and payments for mineral rights.

RATIONAL EXPECTATIONS MARKET EQUILIBRIUM

Distinguishing between the separate influences of supply and demand in market data is complicated, as noted above, if producers form expectations about future prices on the basis of information that determines future demand. The rational expectations approach introduced in chapter 4 provides a method to identify supply behavior and to bring uncertainty explicitly into the analysis. This section provides a brief description of the approach applied to resource supply analysis and discusses the major strengths and weaknesses of the methodology.

Description of the Model

To simplify the exposition, we ignore reserve additions and focus on the market for extracted output. On the supply side, behavior is described in terms of cost functions like those given by equation (2) in chapter 4. These functions are quadratic in decision variables (such as the extraction rate) and linear in stock variables (such as reserves) and exogenous stochastic cost "shocks." In this prototypical model, decisions depend linearly on stocks, expected prices, and the cost shocks. A critical assumption is that all producers face the same cost conditions, so that aggregate supply functions have the same form and content as individual decision rules.[35]

Demand is represented by linear decision rules that depend on current and expected future prices, on stock variables such as energy-using capital, and on various exogenous influences such as consumer income or taxes.[36] These decision rules can be derived from a behavioral model that assumes linear-quadratic utility or production functions.[37]

A novel feature of the rational expectations market model is the characterization of expectations. In contrast to the discussion in chapter 4, where the individual producer faced an exogenous, autoregressive price path, prices at the market level are determined by the interactions

35. This assumption is not required in the theoretical development of the model, as indicated by the analysis of heterogeneous costs and decision rules in Epple, Hansen, and Roberds (1983). In practice, however, fairly strict homogeneity assumptions are required for estimation, given the aggregated nature of available data.

36. In a full general equilibrium model, income is endogenous, and the exogenous influences are tastes, production technologies, population, and the like.

37. See, for example, Eckstein and Eichenbaum (1983a).

between demand and supply decisions and thus depend on the determinants of these decisions. Under the assumption that decision makers are rational in using all available information to form expectations, the interactions between demand behavior and supplier expectations must be addressed. The rational expectations approach to market behavior assumes that firms recognize the interrelationship of capital stock decisions, resource demand, and the market price, and use this information to derive price forecasts that are consistent with the fundamental determinants of market equilibrium.

Specifically, the approach assumes that each resource supplier makes forecasts of aggregate resource *supply*, of energy-using capital stocks, and of other influences on aggregate resource demand, and combines these forecasts with full knowledge of the resource demand function to derive expectations of future resource prices. Thus, supply behavior in the model is hypothesized to depend on demand-side influences such as capital stocks and income, which enter through their effect on price expectations, as well as on supply-side influences such as reserves.

Each firm's conjecture about aggregate supply also is assumed to be correct. This consistency condition, and the assumption that suppliers know the form and content of the demand function, play a critical role in determining the conditions for identification. Sargent (1981) shows how the consistency condition for expectations creates cross-equation restrictions on parameters and disturbances of the supply and demand equations that can be used to identify both demand and supply. The disturbance terms themselves are derived from underlying hypotheses about the cost and demand shocks in the model. These disturbances generally have complex lag structures, in contrast to the simpler specifications typically encountered in least squares models.

Strengths and Weaknesses of the Approach

By focusing on market interactions and expectations, the rational expectations approach can be used to examine how supply responds to structural changes in resource demand and market conditions. For example, the hypothesis that firms know the structure of demand can be employed to ascertain how aggregate energy supply responds to structural changes on the demand side, such as a permanent increase in the productivity of energy-using capital that reduces energy demand or to the introduction of substitutes. In each case, the structural changes in demand will lead to systematic alterations in the coefficients that measure the supply response to price and other economic incentives. Such structural changes in supply that are induced by changes in demand may partly explain why energy supply models with fixed expectations have performed so poorly in predicting supply behavior under changing market conditions in the 1970s.[38] As we will show later, the rational expectations approach also can be used to analyze responses to changing regulatory and institutional constraints.

The weaknesses of the approach were summarized in chapter 4, though they become even more serious at the market level of analysis. Specifically, the assumption that each individual supplier knows the structure of aggregate demand and forms consistent conjectures about the market is extremely strong. However, there are as yet no practical alternatives that incorporate less stringent assumptions. The underlying certainty equivalence methodology also ignores the influence of price and cost variances in forming expectations, which, as Pindyck (1980) shows, can have a substantial (and nonlinear) impact on market price and output paths when resource demand is nonlinear.[39] Finally, the assumption of identical individual decision units is a substantial limitation, as indicated previously.[40]

PRICE AND OUTPUT CONTROLS

Government intervention in the market to control prices or supplies alters market incentives and thus causes a structural change in supply behavior. Except in the most stylized specifications of resource supply behavior—such as the simplest extraction model based on the "*r* percent rule"—few unambiguous qualitative conclusions about the effects of government intervention can be drawn. The conclusions depend on the control regime in place (for example, fixed versus flexible price ceilings); the magnitude of depletion effects and the structure of costs; and the impact of controls at different stages in the supply process (for example, the extent to which exploration as well as extraction incentives are blunted). However, some general observations can be made to illustrate both the conceptual ambiguities and the salient issues for policy analysis.

The first observation is that government controls will alter current supply activities by changing expectations about future earnings. Price controls that are perceived to be permanent (or long-lasting) will tend to shift production to the present, even if they are not a binding constraint

38. See Lucas (1976) and Sargent (1981) for further amplification of this remark in a macroeconomic context.

39. Using equations (15) and (18) in Pindyck (1980, pp. 1209, 1211), it can be shown that uncertainty about both reserves and demand tend to produce more rapid output declines and price increases over time when demand is concave or unit extraction cost is convex in reserves. Only if both unit cost and demand are linear will the certainty-equivalence result obtain.

40. For examples of efforts to relax this assumption, other than Epple, Hansen, and Roberds (1983), see Sargent (1979, chapters 13 and 16) and the references cited therein.

on the current price.[41] The results will be similar for controls on the rate of extraction. Thus, the effect of government controls cannot be ascertained solely on the basis of whether or not they are currently binding.

Expectations about changes in controls will also alter current behavior. Specifically, an expected change in government controls will not have the same effect as an unanticipated policy shift, since the impact of the change already will have been integrated into current production plans. More generally, the degree of myopia about the change and the rate of adjustment by producers to a new control regime have an important influence on the outcome (Pindyck, 1981b).

Another general observation is that controls will mask, but not eliminate, market interactions. To illustrate this point, consider the "prorationing" controls implemented by most petroleum-producing states in the United States during the late 1920s and early 1930s as a conservation measure to limit overproduction in the presence of the common property externalities. Prorationing operated by assigning a production quota, or "allowable," to each producer related to the regulatory agency's forecast of aggregate demand. The production quotas were adjusted according to changes in the demand forecast.[42]

Although market demand prorationing and related constraints were nominally directed at the common property problem, they also led to substantial stability in market oil prices over time. If it is assumed that prorationing authorities acted *as if* to stabilize the price, then it is tempting to conclude that the price can be treated in empirical analysis as being determined by regulation and exogenous to the market rather than by demand. If the assumption is correct, moreover, then empirical estimates of how supply varied with price during the era of prorationing would convey little information about underlying determinants of supply behavior; estimated price responses would be expected to differ widely from responses in the absence of controls.

However, market supply and demand decisions under prorationing are interrelated to the extent that changes in production quotas reflected efforts by regulatory authorities to match changes in supply with changes in demand. The focus of the interactions shifts from the market price to the production quota as the endogenous variable that reflects demand conditions. It follows that a complete analysis of prorationed market behavior requires an explicit hypothesis about the behavior of the regulatory agency as well as assumptions about how interactions among demand, supply, and regulatory constraints determine observed prices and quantities.

41. Lee (1978, 1979, 1981) investigates the implications of price controls in a simple Hotelling model.

42. McDonald (1971) contains a thorough description of prorationing and other conservation regulations.

The remarks concerning government intervention also apply to changes in market conditions such as technical innovation and the introduction of substitute goods. As Lucas (1976) and Sargent (1981) argue in a macroeconomic context, the rational expectations approach to identifying market interactions can be used to distinguish among supply decisions, institutional and regulatory constraints, and expectations of changes in these constraints.[43] Specifically, they argue that imperfectly predictable changes in institutional and regulatory constraints can be modeled as stochastic processes, like the stochastic price and cost processes described in chapter 4. In contrast to transient fluctuations in policy variables, changes in the entire policy regime (such as elimination of the depletion allowance) can be modeled by assuming that producers immediately adapt to the new regime. In this case there is a systematic structural change in the decision rules of individual producers, and thus in aggregate supply responses as well.

However, the rational expectations method in its current state has drawbacks in this regard. Expectations about changes in regulatory policy variables are included but prior expectations about structural changes in the regulatory regime itself are not (though decision makers are assumed to adapt immediately to these changes). As indicated in chapter 4, incorporating these latter expectations requires some sort of adaptive learning model, and analytically tractable models of this type have yet to be developed.

NONCOMPETITIVE MARKETS

Concern over monopolization of nonrenewable resources is longstanding but became especially acute after the oil market disruptions in the early 1970s and the rise of the Organization of Petroleum Exporting Countries (OPEC) as a force in this market. As a consequence, numerous theoretical and empirical studies have been undertaken to learn how departures from perfect competition affect nonrenewable resource prices and outputs.

To the extent that actual conditions in nonrenewable resource markets depart significantly from the competitive ideal, the competitive framework for analysis described in previous sections will yield misleading conclusions and predictions. Ultimately, the magnitude and significance of departures from competitive conditions in resource markets are em-

43. See also Eckstein and Eichenbaum (1983b), who present a rational expectations petroleum market model with an explicit hypothesis about setting of production quotas to stabilize the market price.

pirical questions. However, to address these issues empirically requires a conceptual foundation that makes possible the separation of market structure influences on observed behavior from other forces. In this section, we briefly survey theories of resource markets with noncompetitive suppliers, discuss gaps in these analyses, and examine practical problems encountered in applying the theories. The general conclusion is that, although the conceptual studies suggest strong differences between competitive and noncompetitive behavior, precise inferences are hampered by conceptual and practical limitations of existing models.

At the most basic level, addressing the structure of markets and the actions of an international institution with potential market power such as OPEC raises a number of thorny problems related to market boundaries and the scope of empirical analysis. To begin with, the question of how market interactions affect domestic supply and price is considerably more complex when the domestic resource industry is viewed as part of a world market. If industry output is a small share of the world total, it may be acceptable to view price as exogenous to the industry. If, on the other hand, domestic industry output is a significant share (as may be argued, for example, in the case of U.S. petroleum production), then market interactions on a world scale must be considered in separating demand and supply responses. In either case, forecasts of price and supply from an econometric model would require an assessment of underlying influences on world supply and demand and how they interact. The extent to which regulatory constraints (such as quotas) have segmented domestic and world markets must also be considered.[44] These problems arise in both competitive and noncompetitive environments but are especially acute in the latter.

Analyzing the world market also requires a hypothesis about the behavior of actors such as OPEC. The many attempts to frame such hypotheses have had limited success to date. A popular assumption is that OPEC can be explained as a wealth-maximizing cartel, even though members have somewhat divergent interests.[45] However, simple models of wealth maximization do not seem adequate in explaining observed OPEC behavior or apparent fluctuations in the degree of OPEC market power.[46] Economic factors other than wealth maximization, as well as noneconomic factors, all seem to be important.[47] In the absence of better hypotheses, *ad hoc* specifications of OPEC behavior must be used to

44. See Bohi and Russell (1978) for a review of U.S. oil import control policies and a discussion of the changing structure of the world oil market.

45. See Pindyck (1978a) and Hnyilicza and Pindyck (1976).

46. Moran (1981).

47. The papers collected in Griffin and Teece (1982) provide a good introduction to the state of knowledge about OPEC behavior.

predict their actions.[48] Gaps in understanding of OPEC behavior and uncertainties about market boundaries impose analytic limitations that should be kept in mind in the discussion that follows.

Monopoly and Oligopoly Supply Behavior

Models of noncompetitive resource markets often examine the polar opposite of perfect competition: monopolization by a single producer. In monopoly models that ignore depletion effects on extraction costs and additions to reserves, optimal output decisions are described by an "*r* percent rule" analogous to equation (20) in chapter 2, except that marginal revenue replaces the price in the decision calculus. Given an output path along which discounted marginal revenue less marginal cost is constant over time, prices are chosen by the monopolist to equate demand and supply in each period.[49] These simple models have been extended to incorporate both depletion effects on costs and interactions between extraction and exploration decisions. In the extended models, a marginal profit rule analogous to equation (3) in chapter 2 describes the extraction decision, except that marginal revenue replaces price, while exploration decisions are described by analogues of equation (26) in chapter 2.[50]

A general finding in these studies is that the monopolist sets a higher initial price than the competitive market outcome but increases the price more slowly. By the same token, output is lower initially but higher in later periods, so that the monopolist also is a conservationist who depletes the reserve stock more slowly. Monopolization is also found to retard exploration and reserve development in models that incorporate these aspects of supply behavior. There are exceptions to these conclusions, however, depending on what specific assumptions are made about extraction cost and resource demand elasticities.

In practice, noncompetitive industry structures fall between the two extremes of monopoly and competition and are better characterized as Cournot-like oligopolies or as industries dominated by a single large producer with a collection of smaller producers constituting a "competitive fringe." Under the latter assumption, the large producer acts as a price leader who takes into account the supply responses of the competitive fringe. Competition by the fringe leads to a dominant-firm pricing policy—and thus to a market equilibrium—which differs substantially from the single monopoly outcome. Specifically, in addition to the pricing constraint imposed by competition from the fringe, the dominant

48. See Verleger (1982) for an example, and Bohi (1983) for a critique.

49. For references, see Peterson and Fisher (1977, pp. 694–697), Devarajan and Fisher (1981), and Salant (1982).

50. See, for example, Peterson (1978) and Pindyck (1978c).

firm may choose a "limit price" strategy—with prices set below the unconstrained monopoly level—to limit the fringe's new capacity investment, thus protecting its own market share.[51]

A different approach is needed to analyze an oligopolistic market without a dominant price leader.[52] The equilibrium concept typically used is the one described in chapter 3 for common property externalities, where each producer forms rational conjectures of other producer's extraction and investment decisions and of the market price that matches total supply and demand. Each producer takes into account how changes in its output decisions affect the market price path, given its conjectures about others' behavior, but no producer dominates the market in its pricing decision.[53]

Eswaran and Lewis (1981) introduce a novel element in their oligopoly analysis by treating producers' outputs as imperfect substitutes or complements in demand, rather than as perfect substitutes. They show that when outputs are substitutes, the basic "conservationist" conclusions from the analysis of single monopoly behavior carry through, but if the outputs are complements, exactly opposite conclusions are reached. This finding adds a cautionary note to conclusions based on the assumption that firms' outputs are perfect substitutes. However, Eswaran and Lewis also note that the substitutes case (for example, oil and natural gas) seems more common than the complements case. In addition, the assumption of a fixed number of reserve deposits in their analysis necessarily excludes exploration and development on the extensive margin either by existing producers or by new entrants.[54]

All of these oligopoly models contain strong symmetry hypotheses: oligopolists are assumed to have identical costs and reserves, and competitive fringe producers are aggregated into a single actor. Eswaran and Lewis examine an extension of their model with an asymmetric distribution of resource stocks and costs among producers and find that no unambiguous results relating industry structure and performance can be derived. This is another example of how heterogeneity across individual decision units limits the scope of theoretical market models.

51. See Gilbert (1978). In formal terms, Gilbert uses a Stackelberg leader-follower noncooperative game model, with the dominant firm acting as leader and the fringe collectively representing the follower.

52. See, for example, Aivazian and Callen (1979), and Eswaran and Lewis (1981, 1982). Salant (1976) also studies this industry structure but omits depletion effects; his analysis is based on the "*r* percent rule."

53. In game-theoretic terms, this is the Nash noncooperative game model with so-called simple strategies.

54. Gilbert and Goldman (1978) analyze monopoly behavior under the threat of entry and show that entry threats also induce slower resource depletion and higher prices before entry. However, their model ignores depletion effects and specifies the threat of entry exogenously rather than deriving it as the optimizing decision of actual entrants.

The models reviewed in this section and the competitive market models discussed previously have the same deficiency in their specification of resource demand. In particular, resource demand is assumed to depend only on the current price, whereas the influences of price expectations, capital stock adjustments, and tradeoffs between resource consumption and other goods are ignored. This omission is particularly serious when sharply rising prices due to increased market power provide strong incentives to invest in resource conservation and substitution.

Finally, the open-loop equilibrium concept typically used in models of oligopoly without a dominant firm has drawbacks, as noted in the discussion of common property externalities in chapter 3. Specifically, we noted that an open-loop equilibrium may involve expectations and strategies that are not credible over time, in contrast to "feedback" equilibrium strategies that adjust to changes in reserve stocks and costs. However, preliminary work by Eswaran and Lewis (1982) suggests that this problem may be less serious in oligopoly models with private property resources than in models (competitive or oligopolistic) with common property externalities. A different, and less tractable, "dynamic inconsistency problem" arises in dominant-firm models, as indicated below.

The Dynamic Inconsistency Problem

Recall that in dominant-firm oligopoly models the leader formulates a plan that takes into account its influence on followers in the fringe. The dynamic inconsistency problem in this context is directly related to the way in which the dominant firm is assumed to account for its influence over rivals in formulating its plan. Moreover, the problem can arise even if strategies are in feedback form.

To see how the problem arises, consider the decision problem of a dominant firm and followers at some initial date, with fixed initial stocks of reserves.[55] Given these initial conditions, the followers can formulate closed-loop strategies given consistent expectations about the leader's (and other followers') actions, and the leader can formulate a closed-loop strategy that takes into account the reactions of followers from the initial period onward. In the second decision period, however, initial actions are fixed and both the leader and the followers face new decision problems. Typically, the dominant firm will have an incentive to reformulate its plan by taking into account the influence on followers from the second period onward, rather than from the first period onward. This new plan

55. Our discussion is limited to a brief summary; for details, see Newbery (1981), Roberds (1982), Salant (1982), and Epple, Hansen, and Roberds (1983).

may differ considerably from previously announced intentions. However, if the followers know in the first period that the leader will subsequently deviate in this way, they will put no faith in any announcements by the leader and the hypothesized equilibrium will break down. The problem is avoided only if the leader is obligated by binding contracts to follow its initially formulated strategy. As noted previously, however, binding constraints generally do not exist, particularly in an international context.

Dominant-firm models that avoid the dynamic inconsistency problem can be formulated, but they prove difficult to analyze except in special cases. As might be anticipated from earlier discussions, an important special case in which the analysis is tractable is a linear feedback model with rational expectations, based on linear-quadratic cost functions and linear resource demand curves. Epple, Hansen, and Roberds (1983) present a thorough analysis of this model (in the two-actor case) under a variety of assumptions about market structure (competitive, leader-follower, follower-follower, collusive) and time consistency. A strength of their approach is that it yields closed-form linear decision rules in all of these cases. The authors also show that different assumptions about market structure and time consistency imply different relationships between underlying structural parameters and the parameters of the decision rules. These conclusions provide a basis for empirical tests of hypotheses about the structure and performance of resource markets.

To be sure, caveats mentioned earlier concerning the limitations of a linear certainty-equivalence approach are relevant also in this context. Few results are available for nonlinear models, but Newbery (1981) shows that in the simplest extraction model without depletion effects, time-consistent leader-follower equilibria are closely approximated by follower-follower equilibria. Although further research is needed to clarify connections between dynamic consistency and market structure, Newbery's finding suggests that follower-follower dynamic oligopoly models may be appropriate even where a dominant producer is presumed to exist.

Testing Hypotheses about Market Structure and Performance

As noted above, the linear rational expectations approach described by Epple, Hansen, and Roberds (1983) offers several advantages in empirical analysis of noncompetitive markets, but there are conceptual drawbacks related to the certainty equivalence method.[56] An additional

56. The empirical counterpart to these drawbacks is identified in Epple's (1983) econometric analysis of competitive petroleum supply, which we discuss in chapter 6.

limitation to the approach is that the solution technique for dominant-firm models cannot be readily generalized to incorporate more than two players (in other words, more than one follower). Thus, strong aggregation hypotheses are needed to apply the framework empirically: all nondominant producers must be aggregated into a single "fringe" sector. In practice, the approach may be most applicable to concentrated minerals industries, as in Roberd's (1982) analysis of the nickel industry. In these industries, the number of producers typically is small and aggregation problems are arguably less serious than in the petroleum industry. Data problems also impinge on applications of the framework to petroleum, as indicated in chapter 6.

The simpler empirical methods normally used in studies of industrial organization also have limited applicability because of the inherent dynamics of nonrenewable resource depletion and market interactions. For example, a comparison of the margin between price and (accounting) costs of production is misleading because firms presumably maximize present value rather than current profit. The price can be expected to exceed accounting cost by an amount that reflects the user cost of capital and reserve replacement, even in a competitive market. By the same token, differences between accounting and economic cost measures make it difficult, if not impossible, to use available data in constructing measures of opportunity cost for assessing whether firms are earning supranormal returns.[57] Disparities among discount rates of different firms, because of differences in risk perceptions and preferences, add to the problem of assessing opportunity costs.

An alternative to comparisons of price and cost data involves the use of equity price data to infer actual rates of return. However, inferences about differences between estimated actual and estimated normal rates of return require strong assumptions about the efficiency of capital markets and require data that are generally not available.[58] Yet another possible approach is to assess entry barriers from measures of capital costs and market concentration. The application of this approach is complicated, however, by difficulties in defining industry or market boundaries. Moreover, the presence of entry barriers does not fully describe the conduct of existing industry members.

CONCLUDING REMARKS

The theme of this chapter can be summarized in the observation that empirical analyses of nonrenewable resource supply behavior are prone to serious error when they are based on a theory of individual responses

57. See Stauffer (1971) and Fisher and McGowan (1983) for discussions of this problem.
58. See Broadman (1981).

that is applied with market data. The problem is especially serious in relation to nonrenewable resources because of (i) inherent heterogeneity of resource deposits across both space and time, and thus the heterogeneity of decisions concerned with exploiting these deposits; (ii) complex interactions among the several stages of the supply process and between supply and demand; and (iii) the considerable regulation that nonrenewable resource industries have experienced.

Most models of nonrenewable resource markets contain little more than the theory of individual supply decisions, a rudimentary demand hypothesis, and a market-clearing condition. Moreover, existing models provide few unambiguous indications about the direction and magnitude of the potential error alluded to above. Among the more important remaining gaps are the nature of dynamic interactions between supply and demand, the impact of price and quantity controls, and the effects of price and cost instability. Thus, while theoretical analyses of noncompetitive markets have made considerable progress, conceptual gaps remain and the capability to undertake rigorous empirical investigation of market structure and performance is still at a very rudimentary stage.

To bridge these gaps between theory and application, researchers are forced to supplement theoretical models with *ad hoc* assumptions or to eschew conceptual models in favor of more intuitive and tractable specifications. Although these efforts are necessary to gain practical understanding of resource supply, the divergence between theory and empirical applications raises troublesome questions about the meaning and reliability of empirical findings. The next chaper explores the dimensions of this issue from the standpoint of petroleum supply in the United States.

6
Empirical Methods and Problems

A theory of nonrenewable resource supply provides a general set of guidelines for specifying equations in an empirical model, for determining causal interactions within and among equations, and for choosing among available estimation methods. Unless data and techniques are available to test its implications, however, the theory has no empirical content. To illustrate the difficulties researchers face in empirical applications of resource supply theory, we concentrate on the U.S. petroleum industry and the procedures used to estimate petroleum supply. The U.S. petroleum industry provides a good example of the empirical difficulties, since numerous efforts have been made to estimate the behavior of this industry and the available data are as detailed as those for other nonrenewable resources.

Our discussion begins with a brief statement of practical problems in estimating supply behavior based on the analysis in previous chapters. The material in this section is generally applicable to nonrenewable resources as well as to petroleum supply.[1] We then examine important gaps and inconsistencies between available data on petroleum supply and the information requirements of the theory discussed in chapters 2–5. The chapter concludes with a brief description of empirical models of oil

1. In fact, the problems arise in any study of dynamic economic behavior, but they are especially important in the context of nonrenewable resources, as indicated below.

and natural gas supply, with emphasis on how researchers have tried to come to terms with gaps between theory and application.

EMPIRICAL PITFALLS AND THEIR IMPLICATIONS

Practical problems of specifying behavioral relationships, of aggregating across heterogeneous units, and of identifying interrelated influences arise to some extent in any study of economic behavior. However, they are especially troublesome in analyses of nonrenewable resource supply because of the dynamics of the supply process, the complex structure of resource markets, and the inherent heterogeneity (across space and time) of resource deposits. All of these empirical issues are important, but because of a lack of quantitative evidence on which to judge their implications and interrelationships, no attempt is made to rank them.

Specification Problems

The general implications of model misspecification and of errors in measurement are well developed in the econometrics literature and need not detain us here.[2] We simply note that errors of form, content, and measurement reduce the precision of estimates and the power of hypothesis testing and that combinations of errors may have unknown consequences. Care is needed in setting up an empirical model, but unfortunately the theory offers little guidance on the form of equations and has ambiguous implications concerning their content.

On the question of form, chapters 4 and 5 indicated that a linear model is consistent with the theory only in the special case of linear-quadratic cost functions, linear demands, and additive disturbances, and that there are several deficiencies associated with this specification.[3] More generally, the discussion at the end of chapter 2 showed how the outcomes and predictions of models are sensitive to the choice of specification, and it highlighted the problems in specifying the form of decision rules.

In connection with the content of a supply model, the theory emphasizes the importance of variables that measure the effect of depletion on costs and decisions. Yet, as noted in chapter 2, the use of stock variables such as cumulative production to measure these effects must be regarded as only a first approximation of a more complex relationship.

2. See, for example, Theil (1971) and Schmidt (1976); see also Granger and Newbold (1977) for a discussion of specification problems in time series analysis.

3. See Epple (1983) for econometric evidence concerning the limitations of the linear-quadratic specifications in nonrenewable resource supply applications.

The theory also is not specific about the choice of variables. For example, in measuring depletion at the exploration stage in petroleum supply, there is a choice between cumulative discoveries and cumulative drilling footage. The same problem arises in measuring exploratory effort and development effort where there is a choice of footage, number of wells (perhaps adjusted for depth), or some combination of these measures with other information. As indicated in the next section, these problems are aggravated by measurement errors in existing data and by inconsistencies between data and theoretical requirements.

The specification problem is further complicated by the presence of competing hypotheses and approaches within the theory. A simple example, noted in chapter 2, is the distinction between Peterson's (1978) model with reserve additions as a decision variable and Pindyck's (1978c) model of exploratory effort, which uses a separate production function to relate effort and discoveries. Another example is the contrast between a "decline curve" model of extraction and more complex specifications. There are also practical difficulties in specifying interactions among stages in the supply process. These difficulties have resulted in widely divergent empirical approaches to addressing these linkages, as discussed in the review of petroleum supply models below.

Finally, there is a wide gap between theory and empirical analyses of how uncertainty affects supply. This gap is partly the result of the theory's inconclusiveness about how uncertainty should be represented. The discussion in chapter 4 indicated that uncertainty has complex nonlinear influences on supply responses, with decisions generally affected by the variability of random price and cost fluctuations as well as by average tendencies. In contrast, the certainty equivalence approach commonly adopted in empirical analysis ignores the former influence by focusing entirely on mean prices and costs. This simplification may lead to misleading conclusions, yet the more general theory is too complex to determine the severity of the error that results. The theory also offers few unambiguous conclusions about how risk aversion affects supply behavior and little practical guidance on ways to discern these effects.

Because of the lack of unambiguous predictions in the theory, only the most general conclusions about model strategy and estimation technique can be drawn. One of the more important lessons is that models should contain some stock variables to facilitate the testing of hypotheses about the effects of depletion. Moreover, hypotheses concerning depletion should be distinct from other hypotheses concerning expectations.[4] The linear rational expectations approach described in chapter 4 is desirable in this respect because it is capable of ascertaining the influence of uncertainty and of separating market interactions. Nevertheless, it is

4. For example, Fisher's (1964) model of exploratory drilling includes lagged dependent variables, but these are used as proxies for drilling information and exploration risk.

difficult to test hypotheses about how expectations are formed. Most procedures construct linear decision rules to test the rational expectations formula against some naive extrapolative model.[5] This test does not address the problem of misspecification if responses are nonlinear and the variances of price and cost influence decisions, as discussed in chapter 4.

The lack of specificity in the theory complicates estimation and hypothesis testing about form and content. The theory posits decision rules with lagged dependent variables and complicated error terms involving both autoregressive components and changing variances. In this context, a time-series estimation approach of the type described in Granger and Newbold (1977) and Hansen and Sargent (1980, 1981) is clearly preferable on theoretical grounds to a simpler least squares approach. Yet, the application of time-series methods is limited by the paucity of available data. By the same token, specification tests such as those described in Hausman (1978) can be brought to bear on questions about form, content, and errors in variables,[6] but these tests reveal only the presence of misspecification, not what the correct specification should be.

The specification problems just described are intensified by sharp changes in the economic environment and regulatory regime, such as occurred in oil markets during the 1970s. Both functional forms and coefficient estimates that are "locally" accurate become less reliable when conditions depart substantially from the sample period. Thus, the seriousness of specification problems depends partly on the purpose of the empirical analysis: they are more serious if the purpose is to forecast future supply rather than to estimate historical relationships. To fully understand supply behavior after a structural change, there may be little alternative to waiting until enough data on behavior have accumulated after the change to permit a thorough analysis. The prospect of serious forecasting error can be reduced by employing specifications that incorporate responses to depletion, expectations, and adaptations by producers to changing regulatory and market conditions.

Heterogeneity and Aggregation Problems

The difficulty of deriving aggregate supply responses from disparate individual decisions has been discussed in chapter 5. At a minimum, the classical problem of aggregation bias is encountered: aggregate responses cannot be described as sums or averages of individual responses but depend on the distributions of individual responses by various characteristics. The problem is more serious because of likely differences in the

5. See Wallis (1980) on tests of rational expectations models.
6. Also refer to Plosser, Schwerf, and White (1982) and the references cited therein.

form of individual decision rules and because of perpetual changes in the stock of developed and undeveloped deposits as old sites are depleted and new discoveries are added.

As noted in earlier chapters, differences in the size of reserves and in the response of supply decisions to depletion are an important source of the aggregation problem. Differing responses to depletion reflect disparities in underlying geological and geophysical characteristics of sites and reserves. These differences in turn imply variations in cost conditions among producers and producing regions, differences in input productivity, and differences in the degree to which cost and productivity are sensitive to depletion. For the petroleum industry, several additional sources of heterogeneity can be identified. Crude oil prices vary substantially by type of crude oil, by producing region, and, when price controls are in effect, by output rate and age of discovery. Natural gas prices differed in the past between interstate and intrastate sales and, depending on the price control period, according to depth and age of discovery. Tax rates have varied over time according to the size of the firm and the rate of extraction; regional variation has resulted from differences in royalties and severance taxes. Conservation regulations on production (for example, controls on well spacing and output rate) have varied by state, age of well, extraction rate, and other physical characteristics of the deposits. In each instance, industry aggregates or industry averages ignore firm-specific and site-specific influences on individual decisions, and aggregate behavior is determined by the composition as well as the average of these influences. The differences can be ignored, as noted in chapter 5, only if individual response coefficients are identical for all firms. This assumption imposes implausibly strong restrictions on individual costs, technologies, and regulatory constraints.

The problem is aggravated by changes in the nature of heterogeneous influences over time. This issue is particularly acute in connection with abandonment of old deposits and discovery of new reserves. In general, aggregate behavior depends on the underlying distribution of deposits by size and physical characteristics and on the sample distributions of sites being discovered, developed, and operated at different times. These distributional characteristics are irrelevant only if abandoned deposits are replaced at identical exploration, development, and extraction costs, but these assumptions are tantamount to ignoring the fundamental characteristics of nonrenewable resources.[7]

7. This observation also raises questions about the adequacy of cumulative or stock variables as measures of depletion in an aggregate context. These measures may be acceptable at the individual level, but they ignore the distributional characteristics referred to in the text.

Efforts to deal with the heterogeneity problem in petroleum supply studies are limited to the use of partially disaggregated data by state or region to ascertain cross-sectional differences in the form and content of supply functions.[8] These data have been used in two ways: in a "dummy variables" or "error components" model, where cross-section and time-series observations are pooled; and in separate supply models for different regions (such as onshore and offshore areas) where it is believed that there are significant differences in the form and content of supply functions.[9] Pooling procedures can be used to investigate shifts in the intercept term of supply functions (and to increase the number of degrees of freedom available in the model).[10] As indicated in chapter 5, however, the heterogeneity problem generally transcends the intercept term by affecting other coefficients and, more generally, the form and content of the model. These disparities are not captured by a dummy variables or error components approach, indicating the advantage of estimating separate supply functions for regions where disaggregated data are available. However, the paucity of time-series data places severe limitations on this procedure.

Separating Interdependent Influences

Chapter 5 highlighted a fundamental gap between models of individual and aggregate supply behavior that arises because determinants that can be regarded as fixed at the individual level become variable in the aggregate. Among these factors are various scale effects on cost and input productivity. In the petroleum industry, for example, the probability of success in exploratory drilling and the average size of a discovery depend on the aggregate level of activity, since aggregate behavior changes the distribution (by size and location) of deposits remaining to be discovered.[11] Scale effects also appear in markets for specialized capital goods such as rotary drilling rigs, because lags in the development of new rig capacity mean that rental rates rise with an increase in drilling.

More generally, dynamic interactions appear among stages in the supply process and between supply and demand at the final output stage.

8. MacAvoy and Pindyck's (1975) study illustrates the use of these data.

9. The American Gas Association (1977) Total Energy Resource Analysis (TERA) model is an example, in which onshore and offshore exploration are predicted by two distinct models.

10. See, for example, Pindyck and Rubinfeld (1976) on the application of models using pooled data.

11. Fisher (1964, p. 6) uses this argument to justify an intuitive model of petroleum exploration at the industry level and why the model does not attempt to duplicate the decision process of an individual operator. As noted in chapter 3, common property and informational externalities also create a distinction between individual and aggregate behavior.

The problems encountered in separating these interactions also were discussed in chapter 5. In particular, the standard approach of imposing exclusion restrictions to identify these separate influences may be inappropriate because of interactions between decisions and expectations at the market level.[12] As noted earlier, the rational expectations approach can be used to resolve this problem, but there is a practical concern in addition to the conceptual drawbacks of the methodology. The rational expectations approach to identification is based on maximum likelihood estimation with nonlinear cross-equation restrictions on parameters, as implied by the expectations hypothesis. Application of this method to nonrenewable resources requires a large amount of data to achieve reasonably efficient estimates, an amount that typically exceeds what is practically available.[13]

Further difficulties are encountered in separating interactions at the exploration and development stages of supply. In the petroleum industry, for example, it may be inferred that oil price shocks and foreign expropriation substantially increased the value of U.S. domestic reserves, even in the presence of price controls.[14] Yet, the ability to ascertain changes in the value of reserves is hampered by the absence of direct observations on transaction prices for reserves. As indicated in chapter 5, various measures of these valuations can be constructed, but they all have limitations. In the absence of transactions data, there is no choice but to follow a partially reduced form procedure that essentially treats all producers as integrated firms and imposes an equality between demand and supply in all but the extraction stage of supply to eliminate unobservable variables. Nevertheless, it is questionable whether this approach can pick up changes in the valuation of reserves and thus whether a partially reduced form model that fits historical data will also perform adequately in the future.[15]

Finally, nonrenewable resource industries have been subject to a variety of shifting regulatory constraints and changes in market structure. In the U.S. crude oil industry, for example, domestic prices before 1973 were regulated by state production controls, while import restrictions effectively segmented the U.S. market from the world market. Changes in

12. See Fisher (1966) on the general issue of identification and the application of exclusion restrictions; see Sargent (1981) on their limited applicability in dynamic models.

13. The problem of data scarcity is magnified by the heterogeneity issue, as indicated above.

14. The crude oil entitlements program may have blunted the effect of price controls, at least temporarily.

15. The problem is compounded by the effects of oil market turmoil during the 1970 s on risks faced by domestic producers. In the absence of direct observations on discount rates and other measures of risk aversion, the effect of these structural changes on exploration activity cannot be identified.

domestic production during this period had a direct impact either on the domestic price of crude oil or on the prorationing quotas that stabilized the domestic price. After 1980, on the other hand, domestic oil prices were deregulated and import controls no longer existed, forcing domestic crude oil to compete in a larger world market. Thus, changes in domestic supply became relatively less important in their impact on the domestic price but still had an influence. Between 1973 and 1980, import restrictions no longer existed and state production controls were ineffective, but price controls at the federal level altered the connection between domestic output and domestic price.[16] This was a period of transition from a closed domestic market and rigid price controls to an open market and no controls. At the same time, the structure of world oil trade was undergoing change, with a relative decline in the role of major integrated companies and the growth in OPEC influence.

Regulations on the field price of natural gas sold in interstate markets started in 1954 with only minor adjustments in the price until partial deregulation started in 1978. Throughout much of recent history the interstate natural gas market was in disequilibrium because the price could not adjust to excess demand. As a consequence, changes in natural gas supply affected excess demand but not the market price. Prices in intrastate markets, on the other hand, were market-determined until 1978, when intrastate sales were regulated in the same way as interstate sales. The distinction between interstate and intrastate pricing before 1978 created important distortions on both the supply side and the demand side of the market. Where possible (for example, some offshore gas had to be sold in the interstate market), gas supplies moved from interstate to intrastate markets to earn higher revenues, while important industrial consumers located facilities in gas-producing states to gain access to supplies. Thus, even the market-determined price in intrastate markets was not free of bias due to regulatory policy. Regulations also had significant effects on the demand for natural gas, with increased prevalence of interruptible contracts, dual-fired capability, and other devices to promote substitutability to limit the risks and costs of curtailment. Regulatory distortions have persisted in the form of long-term contracts that limit price flexibility in the face of changing market conditions.[17]

These institutional and regulatory changes in markets for oil and natural gas introduce another level of complexity into the relationship

16. Montgomery (1977) thoroughly describes the institutional structure and the economic implications of oil price controls.

17. See Russell (1972) and Huntington (1978) on this aspect of gas price controls, and Broadman and Montgomery (1983) for a survey of regulatory and contractual issues in the gas industry.

between domestic supply and price. As noted in chapter 5, the common simplifying assumption of an exogenous oil price during the prorationing period is invalid if changes in demand fed back to changes in production quotas, which in turn fed back to supply decisions. In this case, market interactions expressed through changes in quotas must be taken into account if the analysis is to shed any light on the underlying determinants of supply behavior. Addressing these interactions requires an explicit hypothesis about the regulatory process and its effects on supply decisions.

Further to this point, if expectations about regulatory constraints are part of producers' decisions, the effect of changes in constraints will depend on the extent to which these changes are anticipated and how rapidly producers adapt to changes once they are in place. Models that fit the data well during past or present regulatory regimes may perform poorly in predicting supply responses under different regimes.

This concludes the general review of practical problems of nonrenewable supply analysis. In the balance of this chapter the focus is on U.S. petroleum supply, beginning with inconsistencies between theory and data.

PETROLEUM SUPPLY DATA

The most serious discrepancies between variables defined in the theory of nonrenewable resource supply and observed measures of petroleum behavior stem from the fact that the theory is concerned primarily with individual decisions of the firm, whereas the data reflect market interactions and aggregation over heterogeneous individuals. There are also a number of specific inconsistencies between what is required and what is available. In describing these pitfalls in the data, we do not attempt to review all relevant data on petroleum supply or to provide an in-depth description of those data that are discussed.[18] Instead, our purpose is to illustrate the more general problems confronted by empirical researchers. The section is in three parts. The first pertains to the definition and measurement of reserves, the second to deficiencies in cost and input data, and the third to aggregation problems.

Concepts of Reserves

Differences in how petroleum reserves are defined in theory and measured in practice go to the heart of the inconsistency between theory and application. As indicated by the discussion of depletion effects in chapter

18. MacAvoy and Pindyck (1975) provide a useful catalog of petroleum supply data and sources.

2, the theory requires a measure of reserves that are physically present in the deposits discovered, developed, and operated. At the extraction stage, this physical measure of reserves is related to reservoir pressure and the rising cost of extraction as the reservoir is depleted; at the exploration and development stages, this measure is related to the rising cost of finding new deposits as the stock of cheaper, undiscovered resources declines.

This theoretical concept contrasts sharply with the definition of "proved" reserves and new discoveries found in petroleum supply data. Measures of proved reserves are defined as volumes of resources that are estimated "with reasonable certainty to be recoverable in future years from known reservoirs under existing economic and operating conditions" (American Petroleum Institute, 1976, p. 1). Discoveries are defined as additions to proved reserves in new fields or pools; extensions are proved reserve additions in existing fields and pools; and revisions are residual changes in proved reserves on the basis of new information gained from exploration and operating experience.[19]

The quoted phrase in the previous paragraph highlights both the differences between theoretical and practical concepts of reserves and the pitfalls of attempting to combine the two concepts. To begin with, proved reserves data are measures of ultimate resource recovery which reflect projections (by those who produce the data) of future production profiles from known reservoirs. Thus, data on proved reserves measure the inventory of current and future production opportunities rather than the physical size of reserves in place.[20] As a consequence, the relationship between reported data and the depletion effects discussed in chapter 2 is far from clear.

Even more troublesome, from a practical standpoint, is the fact that when calculations of proved reserves are based on estimates of cumulative future production, they implicitly involve hypotheses about extraction from existing reserves (typically a "decline curve" assumption). This procedure raises serious questions about the applicability of reserves data in *any* model of extraction behavior in which reserves are hypothesized to influence production through their effects on cost and capacity. Specifically, it is difficult, if not impossible, to distinguish between the hypotheses in the model that relate reserves to production and the hypotheses embedded in the data that relate production to reserves. This problem is somewhat less serious in analyses of new discoveries; yet, the use of data on proved reserves and discoveries requires assumptions that rigidly link

19. See American Petroleum Institute (1976) for more precise definitions.

20. This point is argued in greater detail by Lovejoy and Homan (1965). Data on "original oil in place" also are available; however, these data are derived by applying an average historical recovery factor to data on proved reserves and thus have the same drawbacks.

exploration and extraction. Moreover, the assumed linkages must be consistent with those that underlie the data if measurement and specification errors are to be avoided.

There are also measurement errors in the data on proved reserves.[21] These data are based on existing economic and operating conditions, so changes in reported reserves and discoveries will differ from actual supply responses when these conditions are altered. For example, subeconomic reserves already discovered are not counted as proved reserves but will be added to proved reserves at higher prices, and the reserve additions resulting from the reinterpretation of subeconomic deposits cannot be separated from additions resulting from new discoveries. This revision of measured reserves with changing conditions raises difficulties in separating demand and supply responses and in measuring responses to expectations about prices and costs.

Moreover, the conservative requirement of "reasonable certainty" in the calculation of proved reserves imparts a downward bias in the relationship between actual reserves and economic incentives. Proved reserves may understate operators' actual expectations of recoverable reserves, and the degree of understatement will be affected by price and cost conditions. Another measurement problem arises from the lack of crediting extensions and revisions back to an original discovery year, an omission that raises serious problems in sorting out depletion effects at the aggregate level.

The incompatibility of reserves data with existing theory, and the need for modifications of the theory to account for the characteristics of available data, should be clear from the discussion above. Yet, the modifications needed in the theory are by no means apparent. One possible direction is to reformulate the theory with a focus on how inventories of reserves influence extraction decisions and to reformulate exploration and development decisions explicitly in terms of capacity. However, the problem of unraveling the hypotheses of the model from those used to construct the data remains to be addressed.

Costs and Inputs

Cost data are scarce and, as might be expected, are reported only for broad categories of expenditures. Measures of incremental costs or of the relationship between costs and reserves simply do not exist. Moreover, the aggregated cost data are subject to significant sampling errors and do not distinguish the effects of technical change from other factors.[22]

21. These problems are reviewed in Lovejoy and Homan (1965).
22. See Fisher (1964, pp. 53–63) for a discussion of these points.

The only available input data reported with any detail concern exploratory and development drilling. Observations are available on footage drilled, distribution of depths drilled, and total number of wells (oil, gas, and dry holes). Several shortcomings are encountered in applying these data. To begin with, total footage, number of wells, and average depth do not provide unambiguous measures of exploratory or development effort. Footage measures the scale of activity, whereas the number of wells measures the intensiveness of input use. Adjustment for well depth creates an additional problem of distinguishing between changes in depth due to depletion and changes due to alterations in the scale or intensiveness of activity.

The absence of a clear measure of industry effort leads to classic problems of omitted variables and measurement errors. The problem is particularly acute in distinguishing between activities on the intensive and extensive margins, since data on capacity, improved recovery inputs, and total acreage explored and developed also are scant. A further difficulty arises in separating exploration activity from development activity. This difficulty is indicated by the fact that information about dry holes is recorded at the development stage.[23] At a more general level, information on the underlying physical distribution of petroleum deposits is limited, and observations on operators' beliefs about this distribution are unavailable. Consequently, *ex ante* probabilities of making a discovery and the size distribution of discoveries cannot be estimated directly. As noted in chapter 4, expectations must be inferred from other hypotheses based on observable information.

Heterogeneity, Market Interactions, and Regulations

Partially disaggregated data (by state and region) are available for output, reserves, and drilling activity. These data are an imperfect substitute for measures that are disaggregated by underlying geophysical conditions, though they may be used to examine broad contrasts among supply responses. Another problem is aggregation over time. Neither output data nor reserves data are adequately correlated to an original discovery year. Consequently, the assessment of depletion effects is complicated if not impossible, particularly on the extensive margin of new discovery and development. Data on cross-sectional price and quality differentials also are limited; discount rates are not directly

23. A related problem concerns the use of data on geological and geophysical activities such as seismic mapping. The output of these activities is information that is not readily quantifiable on the size and location of deposits.

observable and must be inferred from observable interest rates and other financial data.

Difficulties in identifying market interactions and the effects of regulatory constraints also reflect deficiencies in the data. Information on energy-using capital stocks is highly imperfect and unreliable for distinguishing demand responses, and the shadow prices of new discoveries and reserves are not directly observable. The joint-products nature of oil and natural gas poses special problems in the absence of knowledge about gas-oil tradeoffs on the supply side and substitution possibilities on the demand side.

Shifting regulatory constraints insert another dimension of measurement error. Specifically, price controls create a discrepancy between an incentive price for exploration and extraction and the observed price. Prorationing controls on production mask the responsiveness of supply to price. Import controls and depletion allowances introduce further distortions. Finally, data on the number of producers exploiting affected sites and on the timing of entry and exit decisions are needed to assess producer responses to common property externalities, but little of this information is available.

The comments in this section highlight the serious limitations of the data used in empirical analysis of petroleum supply and illustrate the problems of bridging gaps between the data and the theory of nonrenewable resource supply. The next section examines how analysts have approached the task of estimating petroleum supply behavior in the face of these difficulties.

CHARACTERISTICS OF PETROLEUM SUPPLY MODELS

Two fundamentally different approaches have been used to estimate petroleum supply—labeled here as optimization models and econometric models—with equally distinct advantages and disadvantages in their use of the theory and data and in the nature of the information conveyed.[24]

24. These terms are potentially misleading because optimization models may include econometric elements, and many econometric models are explicitly derived from a theory of optimizing behavior. To avoid confusion, optimization models are sometimes referred to as process models, but the latter terminology is misleading in the present context. Many process models focus primarily on descriptions of the geological and technological aspects of resource supply, whereas this study focuses on models in which these aspects of supply play a secondary role in the derivation of economic decision rules. For this reason, we avoid the terminology of process model and do not include in our review of supply models the estimation procedure developed by Kaufman (1963) and implemented by Eckbo, Jacoby, and Smith (1978). Nevertheless, a combination of process and economic models has considerable potential for increasing practical understanding of nonrenewable resource supply, as indicated in chapter 7.

Our objective in describing the alternative methodologies is to focus on general characteristics as they relate to the theory of resource supply and the estimation problems discussed above.[25]

Optimization Models

Optimization models attempt to solve the present value problem posed by theory directly. A computer program is used to calculate decision paths that maximize the stream of discounted net income for a given set of constraints and parameter values.[26] The program may be reiterated for alternative parameter values and alternative constraints to determine how price changes, policy changes, and other factors affect the calculated outcome. The clear advantage of this approach is that complex objective functions can be used to calculate supply responses without concern for deriving estimable decision rules that must be fitted to historical data.

The optimization models of supply are completely deterministic—all information required for optimization is assumed to be known with complete certainty by suppliers—although they may be supplemented with econometric demand models that include random error terms. The conceptual implications of this simplification, particularly as related to uncertainty about future costs and prices, were addressed in chapter 4 and need not be repeated here.[27] As an empirical matter, the disadvantage of this approach is that it provides no objective way of testing hypotheses about behavior. All ingredients in the models enter as assumptions. The models yield predictions about supply activity, and these can be evaluated in relation to actual outcomes. However, the reason for success or failure cannot be tested. Consequently, the predictions may appear to be successful for limited periods even though the objective function is misspecified, assumptions about parameters and constraints are erroneous, or assumed values for future prices and costs are incorrect. On the other hand, if the predictions are unsuccessful, there is no way to locate the source of the problem.

Individual optimization models differ in many details, including the definition of net income; the range of decisions that encompass exploration, development, and production; and the richness of policy constraints. However, all models in this category employ a similar

25. For a detailed comparison of assumptions, procedures, and results for ten prominent models, see Clark, Coene, and Logan (1981).

26. Examples of optimization models include: American Gas Association (1977, 1979) TERA Offshore model; Adler and coauthors (1979) DFI model; U.S. Department of Energy (1978, 1979a, 1979b); Kim and Thompson (1978); and Adelman and Paddock (1978). An application to coal supply is given in Modiano and Shapiro (1980), and to bauxite supply in Pindyck (1977).

27. In particular, the problems of decision making under risk and uncertainty are eliminated as a result.

characterization of the supply process. In general, there are no dynamic interrelationships among the three stages of the supply process. Instead, the decision process is completely sequential—extraction depends on development, and development depends on exploration—with no allowance for feedback among stages. A common specification in models that encompass all three stages is to fix production and development activities in strict proportion to exploration. The procedure is employed to make the optimization problem tractable, but its effect is to place the focus of attention entirely on the exploration stage. The results of the exploration model completely determine the results for development and extraction. Because of this simplification, optimization models that encompass the entire supply process contain little more substance than less ambitious models that are concerned only with exploration activity and new discoveries.

Optimization at the exploration stage is determined primarily by the specification of a fixed relationship between the rate of investment in exploratory activity and the volume of new discoveries. The specification of this relationship, in turn, requires two essential pieces of information: the value of discoveries and the volume of undiscovered resources. The unit value of new discoveries is calculated from a profile of exogenous future prices and production, both of which are incorporated into the models as if they are known with complete certainty.[28] The volume of undiscovered resources is generally obtained from published estimates made by the U.S. Geological Survey, the National Petroleum Council, and the Potential Gas Committee. These estimates are used to calculate the volume of new discoveries that will result from an additional unit of exploratory effort. The volume of new discoveries is multiplied by the unit value of new discoveries to determine the total value of assets in the ground and, at the same time, to determine the return on investment from exploiting these resources.

Two alternative procedures are used to calculate the volume of new discoveries. The most common approach uses a "finding rate curve," which is a functional relationship between the amount of discovery per unit of effort in a geographic area and the cumulative amount of total exploratory effort in that area. The finding rate curve is a monotonically decreasing function of cumulative exploratory effort, indicating that marginal productivity of new investments declines with depletion. The slope of the curve is fixed by assumption and is generally buttressed by past performance, revealed through fitting semilogarithmic curves to actual data. The position of the curve (in other words, the intercept) is

28. Note that the assumed production profile used to calculate the value of discoveries presupposes the subsequent development and extraction decisions.

determined from the separate estimate of undiscovered resources referred to above. The position is set such that the area under the finding rate curve equals the total volume of discovered and undiscovered resources.

The alternative approach for calculating the volume of new discoveries is to assume that exploratory effort is uniformly successful in finding new resources—in other words, a constant success ratio for exploratory wells drilled—but that the size of discoveries declines with the cumulative amount of exploratory effort. This procedure requires information about the rate of decline in the size of discoveries as well as an estimate of the total number of discoveries of each size (and hence cumulative resources) to be found. This information is generally determined from extrapolations of historical experience.

The measure of exploratory effort is a crucial yet troublesome element in the calculations. As indicated in the previous section, available data do not adequately capture the variety of effort involved, nor do they provide an unambiguous measure of exploratory effort. The common measure of effort employed in the models—footage drilled—does not take into account prior geophysical investigation, well depth, and other features that account for differences in costs. The margin of error is reduced if the model is applied to regional rather than national data, including a distinction between onshore and offshore areas. Nevertheless, no single measure of effort can be expected to capture the diverse aspects of exploration.

Another general characteristic of the optimization models is the crude mechanism by which exploration decisions are spread out over time, if in fact the tradeoff between investment today relative to investment tomorrow is incorporated at all. In some models, the tradeoff depends entirely on the presumed course of future prices. Unless the price path rises fast enough, the optimal solution is to lump all investment in the current period, as determined by the marginal point on the declining finding rate curve (or its equivalent). The result is an unrealistic pattern in the timing of new discoveries and, similarly, unrealistic development and extraction profiles that are automatically linked to the volume of new discoveries. Other models are completely myopic about future prices and use a drilling cost constraint to limit the amount of exploratory effort in any single period.

To summarize, the optimization models replicate the present-value-maximization problem posed by theory as a deterministic expression of producer behavior. The decision process is entirely mechanical, and most if not all inputs to the model are determined exogenously to the decision process. Although depletion plays a central role in the calculations, it is represented by simple mechanical relationships that remain fixed despite changes in the economic and operating environment.

With respect to the aggregation issues raised earlier, the optimization models make the extreme assumption that the market behaves in the same way as the individual producer. The form and content of the objective function are assumed to be unaffected by the substitution of market data for measures of performance by an individual firm. Specifically, the methodology implicitly assumes that all resource deposits within the geographical unit of measurement (state or region) are physically identical and that producer responses within each unit are the same. Finally, except for models with a market-clearing condition for extracted output, economic influences are exogenously determined, as at the individual level, rather than endogenously determined as at the industry level.

Econometric Models

Econometric models are statistical relationships estimated with historical data, where the relationships are derived explicitly or implicitly from a theory of nonrenewable resource supply. In contrast to the optimization models, econometric models explicitly include random influences on variables, so that statistical tests of hypotheses can be applied.

In light of the aggregation issues emphasized in this chapter, it is useful to separate the econometric models into two groups: those derived from a theory of individual behavior and those based on intuitive relationships for the industry. Models in the second group, which are the more numerous, do not involve a formal derivation of statistical relationships from a theory of optimizing behavior. Instead, the specifications are *ad hoc* representations of causal relationships that are suggested by theory.[29] By fitting linear (or log-linear) regression equations to historical data, they yield measures of the association between the explanatory variables and the observed measures of exploration, development, and extraction activities. The coefficients of the explanatory variables measure the responsiveness of supply activities to changes in these variables, which may be used to test hypotheses about industry behavior. Under the assumption that the estimated coefficients are stable from the past to the future, the regressions can be used to generate supply forecasts.

The model developed by Fisher (1964) is the seminal work of this type and was followed by a series of other models that employ the same basic

29. Examples are Fisher (1964), Erickson and Spann (1971), Khazzoom (1971), American Gas Association (1973) TERA Onshore model, Spann and Erickson (1973), MacAvoy and Pindyck (1973, 1975), Rice and Smith (1977), and Pindyck (1978b).

procedure with numerous variations. Fisher's model is concerned with explaining the volume of crude oil discoveries indirectly from three separate log-linear equations that explain the number of exploratory wells drilled, the average success ratio, and average size of a discovery. Multiplied together, these three variables equal the volume of new discoveries. Among the explanatory variables, the most important are the price of crude oil and lagged values of the dependent variables. The estimated price elasticities are 2.85 for drilling, −0.36 for average success, and −2.18 for average size, giving an overall price elasticity of 0.31 for new discoveries (equal to the sum of the three components). These estimates are of current interest, even though derived from data for the 1946–1955 period, because of their comparative success in predicting more recent events. The estimates imply that drilling activity responds sharply to changes in crude oil prices, but that the volume of new discoveries does not. These conclusions come closer to predicting events in the last decade of sharply rising oil prices than virtually any of the other econometric models.

Erickson and Spann (1971) modified Fisher's model by separating oil and natural gas discoveries to test the joint-product characteristics of the two resources. They concluded that the price responsiveness of crude oil discoveries is significantly larger than Fisher's estimates (about three-fold),[30] and that the price of gas is not statistically significant in determining supply activities. The latter result is attributed to the lack of distribution pipelines during the sample period (1946–1959) and the consequent lower market value of gas relative to oil.

The most ambitious adaptation of Fisher's basic model is represented by the models of MacAvoy and Pindyck (1973, 1975) aimed primarily at explaining natural gas supply. In addition to the Fisher-like discoveries model, MacAvoy and Pindyck add separate equations for extensions and revisions of reserves, for extraction, and for final demand. The models thus incorporate all three stages of supply and combine supply and demand forecasts to determine equilibrium prices. Although conceptually more satisfying, their models still performed poorly. Estimates of the price responsiveness of new gas discoveries turned out to be unrealistically high, whereas estimates of market-clearing prices turned out to be too low.[31]

30. Eyssell (1978) obtains an even larger estimate of the price elasticity of new discoveries (4.07) by adjusting the measure of average discovery size in Fisher's model for subsequent extensions and revisions.

31. A subsequent version by Pindyck (1978b) eliminates prices from the success ratio and discovery size equations, so that the price response of new discoveries depends entirely on the response of drilling activity.

The intuitive specifications used in these models have been criticized because they cannot be derived explicitly from the maximization of an objective function for individual suppliers. As a consequence, it is argued, the specifications may be subject to error, and the parameters appearing in the equations have no clear economic interpretation. This somewhat looser procedure can be justified, on the other hand, because of the differences between individual and industry behavior discussed earlier. If the industry is influenced in different ways and by different factors than individual behavior, an aggregate specification based on individual optimizing behavior also is subject to specification error. Nevertheless, *ad hoc* econometric models have performed poorly, and the number of variations in the specification already employed in the search for a workable model has virtually exhausted the practical options. It is difficult to avoid the conclusion that a fundamentally new approach to supply analysis is required.

Among the econometric models derived explicitly from a theory of optimizing behavior, two approaches deserve to be distinguished. The first approach applies conventional production theory to the resource sector: the typical firm employs inputs in production to obtain outputs in the form of new discoveries, as in Epple (1975), or in the form of additions to reserves, as in Cox and Wright (1976). In Epple's model, inputs are taken to be labor, capital, and oil-bearing land. Depletion effects do not enter the firm's decision problem—the problem, in fact, is completely static—but nonrenewability does appear implicitly in the specification of oil-bearing land supply. From the production function, demand equations for inputs and supply equations for discoveries are derived and are used to specify the estimating equations. Cox and Wright, in contrast, develop a model more closely related to the dynamic decision theory of chapter 2. Producers are assumed to maximize the present value of after-tax revenues subject to a production function relating reserves and production and an "inverse cost function" that relates investment outlays to replacement of reserves. The latter function is independent of cumulative activity, so that depletion effects enter the model only in the relationship between extraction and reserves. From this model, Cox and Wright derive estimating equations with reserves as the dependent variable.

Both models make a leap of faith—like the optimization models—that the industry can be characterized by the typical firm, without consideration of problems that arise from aggregating over dissimilar firms and heterogeneous resource deposits. Both models also use complicated calculations to derive measures of how price and tax incentives influence activity. The complexity of these calculations makes it difficult to test the responsiveness of behavior to changes in prices, costs, and taxes, and to discern the impacts of dynamic supply-side market interactions. Epple's procedure is similar to the optimization models, in that separate hypotheses about the link between discoveries and production are used to infer the value of discoveries in terms of the price of extracted output. Moreover, in measuring this value, suppliers are assumed to have myopic expectations about prices and costs. Cox and Wright take a similar approach in calculating the value of new reserves.

The specification of causality in the Cox and Wright model is an additional shortcoming. The estimating equation for reserves contains extracted output, an incentive variable related to the price of output, and a market-demand prorationing factor as explanatory variables. Their rationalization for this specification—that prorationing exogenously determines both price and demand—is less than convincing in the light of the discussion in chapter 5, particularly with respect to the potential endogeneity of the market demand factor. Inclusion of all three variables also begs the question of their separate explanatory power under prorationing, notwithstanding their apparent significance.[32]

A more recent econometric approach based on a model of optimizing behavior, introduced by Epple and Hansen (1981) and applied by Epple (1983), uses the linear rational expectations methodology to estimate the underlying decision rules of resource suppliers in an explicitly dynamic framework.[33] Like the optimization models described earlier, the producer is assumed to maximize the present value from exploiting a nonrenewable resource. Unlike the optimization models, however, uncertainty about future prices and costs plays a crucial role in the producer's objective function. The objective function is used to deduce stochastic decision rules that incorporate responses to depletion, given a specification of how expectations about future values of variables are determined. The parameters in the estimating equation thus appear in either the objective function or in the expectations equation, giving a clear interpretation to all parameter estimates.

A number of simplifying assumptions are employed to make the models tractable. The derivation of linear estimating equations requires a linear-quadratic cost function and stationary linear expectations equa-

32. Cox and Wright also aggregate oil and natural gas by constructing Divisia indexes of prices, outputs, and reserves. This procedure raises further problems regarding heterogeneity and identification because the index weights vary over time and depend on market conditions. Evidence of misspecification can be seen in Durbin-Watson statistics, which indicate a lack of serial correlation in the residuals for both differenced and undifferenced models.

33. Roberds (1982) uses the rational expectations approach in developing a dominant-firm model of exhaustible resource supply for application to mineral industries such as nickel.

tions. As Epple notes, the empirical results obtained from the model cast doubt on the validity of the first hypothesis. The second simplification requires linear demand curves. To estimate new discoveries, Epple also employs the simplifying assumptions that extraction is fixed in relation to the volume of reserves and that additions to reserves are fixed in relation to the volume of new discoveries. Thus, the exploration decision rule ignores dynamic interactions with development and extraction decisions. Aside from this simplification, however, the exploration decision rule is behavioral rather than mechanical and may be used to test hypotheses about the path of output and the volume of ultimate recovery.

In estimating his model with aggregate data, Epple also employs an ingenious procedure for crediting production and additions to reserves back to an original discovery year. This procedure plays a crucial role in his derivation of the value of a discovery. In contrast to his earlier (1975) study, Epple also assumes that suppliers rationally forecast changes in output prices and taxes in determining a discovery's value. Unfortunately, however, the practical import of Epple's "vintaging" procedure and the reliability of his estimates are severely limited by reliance on proved reserves data. As indicated above, these data are fundamentally incompatible with the type of conceptual model on which Epple's analysis is based. Reliance on proved reserves data is, in fact, a generic weakness of the models we have discussed.

To date, the rational expectations approach has not established a record of empirical success or failure on which to make a practical evaluation. However, the approach is unique among empirical resource supply models because it incorporates dynamic responses to depletion as an integral part of the model, allows for uncertainty about costs and prices, and posits an explicit mechanism for determining expectations that are consistent with the factors influencing the market. As such, the rational expectations approach preserves all of the essential ingredients in resource supply theory and offers a potentially fruitful alternative to the class of *ad hoc* econometric models. Nevertheless, as in the other models derived explicitly from a theory of optimal individual behavior, the issue of aggregating the model to the industry level for use with available data is a major problem that remains to be tackled.

CONCLUDING REMARKS

Different perspectives on understanding the behavior of petroleum supply—from the standpoint of data, methodological considerations, and specific empirical studies—all lead to the conclusion that the disparity between theory and application seriously impedes understanding of observed behavior. The problem is no less serious, though it may vary in particulars, for other nonrenewable resources. In light of the shortcomings, it is easy to be less than sanguine about the prospects for further advances in knowledge and thus about the usefulness of additional research on the topic. Our view is more optimistic, however. Several avenues of inquiry promise significant increases in understanding, as indicated in the next chapter.

7
Summary and Research Suggestions

We have argued that practical understanding of nonrenewable resource supply is hampered by gaps among theory, methodology, and data. The arguments may be summarized in four general propositions:

- The theory of nonrenewable resource supply, based on self-interested responses by individual producers to depletion effects, current market conditions, and expectations of future conditions, predicts a complex pattern of supply behavior with intertemporal relations among decisions at each stage of the supply process and dynamic interactions among decisions taken at the various stages. Specifically, the theory predicts extraction, development, and exploration responses based on expectations of future prices and costs as well as on current conditions. Responses also depend on stocks of developed and undeveloped reserves, among other influences, as indicators of current capacity and cost. The theory provides a framework for analyzing how activities at one stage are affected by activities at the other stages and how activities at all stages ultimately depend on prices and costs at each point in the supply process. As in the analysis of a single stage, the connections among the stages are inherently dynamic.

  Extensions of the basic theory address numerous topics, including common property externalities, technical change, joint products, and the influence of uncertainty and attitudes toward risk on supply behavior. Taken as a whole, the theory suggests hypotheses about the determinants of observed supply behavior. It provides general, qualitative guidance, but few firm conclusions, on how supply responds to changes in economic influences and regulatory constraints. The theory by itself is inadequate for addressing these issues; careful empirical analysis is required to unravel the dynamics of supply at the market level and to gauge the impacts of uncertainties at each stage of the process.

- Empirical specifications that gloss over the above-mentioned characteristics of nonrenewable resource supply behavior run the risk of seriously misrepresenting the supply process conceptually and of yielding distorted coefficient estimates, inferences, and predictions. However, practical application of the theory is hampered by its inherent complexity and by gaps in the conceptual framework. For example, the theory can be extended to incorporate uncertainty about future prices and costs, but the complicated nonlinear responses predicted by general models limit their empirical applicability. On the other hand, linear stochastic models are based on simplifying assumptions that may or may not be valid and are difficult to test. Another general weakness in the theory concerns strategic supply responses to common property externalities or noncompetitive market forces. In addition, for specific applications, the basic theory must be modified to reflect the peculiar institutional (as well as technical) characteristics of the industries and markets under study.

- The theory is particularly deficient in its primary focus on individual decisions rather than market behavior. Application of the individual decision theory with aggregate data also risks error by ignoring the implications of cross-sectional heterogeneity, dynamic market interactions, and the implications of regulations that overlay these interactions.

- Finally, many important data on supply activities (such as discoveries and reserves) are defined in ways that are fundamentally incompatible with the data requirements for applying existing theory, and perhaps with *any* behavioral model of supply. Addressing this issue requires both a reformulation of the theory toward consistency with available data and a careful investigation of the procedures used to develop the data.

The standard we have set out for analysis of nonrenewable resource supply—achieving consistency among theory, data, and estimation methods—is a strong one, particularly from the standpoint of evaluating existing empirical studies. This standard clearly is not appropriate in all contexts, and blanket criticism of empirical models based on the gaps identified in this study would be misplaced. For example, the drawbacks to optimization models may be perceived as less serious in the context of using the models to aid in corporate planning or government policy analysis, where the purpose is to gain a general understanding of the magnitudes involved when the regulatory milieu and market conditions are changing.

We do believe, however, that our standard is appropriate for highlighting the limitations of existing analysis and for suggesting research directions to relax these limitations. Knowing the limitations is particularly important if, as is unfortunately too often the case, the models function as a substitute for, rather than a supplement to, the sound judgments of decision makers. This problem is especially serious if the models themselves are based on naive descriptions of resource supply like the "*r* percent rule."

Moreover, identifying research questions that advance scholarly understanding of nonrenewable resource supply is entirely compatible with the development of more useful empirical models from a "rough and ready" perspective. In particular, we believe that three research areas—further application of rational expectations methods, analysis of the regulatory process, and evaluation of data—have great potential from both perspectives. More specific recommendations are offered in the next section. On the other hand, we do not want to minimize the work required to bridge the gaps identified in this study. Practical analysis of resource supply cannot await a golden age in which all of the problems are solved, and the value of back-of-the-envelope analysis should not be underestimated given the current state of the art.

DIRECTIONS FOR FURTHER RESEARCH

Our recommendations for additional research focus on studies that might help to bridge the gaps mentioned above. Because empirical studies of nonrenewable resource supply are few in number and narrow in their range of approaches, our judgments are intuitive, and cannot be buttressed by reference to contrasting empirical findings. Conclusions are grouped into four parts pertaining to general specification issues, uncertainty and expectations, market-level analysis, and strategic behavioral issues.

Depletion Effects, Dynamic Supply Interactions, and Reserves Data

At a minimum, theory provides an intuitive guide to dynamic specifications of behavior, with resource stocks and cumulative output (or effort) measuring depletion effects and dynamic interactions among stages of supply. Empirical studies should fully incorporate these aspects by eschewing static specifications and by treating all stages of the supply process as interdependent decisions. The major methodological challenge is to develop specifications that are rooted in the theory—so that its lessons can be applied and its hypotheses tested—and are compatible with available data. An example of such an undertaking is the work of Farrow (1983) and Stollery (1983), who modify the prototypical cost function used in this study to reflect specific technological characteristics of minerals production. Another important part of this effort should be an investigation of the data set itself, following the example of Lovejoy and Homan (1965), so that implicit assumptions made in compiling data can be clarified and their implications for empirical analysis can be discerned.

Expectations and Uncertainty

A significant weakness of existing empirical studies, even those that seem to fit historical data well, is the inability to forecast responses to important changes in market conditions, such as those that occurred in the petroleum industry during the 1970s. One possible explanation for this poor record is the absence of expectations hypotheses that are consistent with the actual determinants of market demand and price. To address this concern, applications of the rational expectations methodology to behavior of nonrenewable resource supply should continue along the lines described in Epple and Hansen (1981) and Epple (1983); furthermore, the applications should be extended beyond one stage of the supply process (such as exploration) to examine dynamic interactions among the stages.

Given the current state of this methodology, further applications initially will take the form of a certainty equivalence approach, with linear model specifications and additive random disturbances. An important subsequent step would be the application of nonlinear rational expectations models, based on the theoretical work of Pindyck (1980, 1981a, 1982), to facilitate testing more general hypotheses. In particular, nonlinear models could be used to assess how the variance of price or cost and various forms of "learning" influence behavior. However, practical use of such models awaits further development of the rational expectations methodology.

Other refinements of methods for incorporating uncertainty would also be useful. Existing rational expectations models hypothesize *ad hoc* linear autoregressive stochastic processes to represent fluctuations in cost and reserves. An integration of this methodology with sampling theories of exploration activity and discoveries, as in Kaufman (1963), Eckbo,

Jacoby, and Smith (1978), and Ramsey (1980), and with specific case studies to help shed light on uncertainties in costs of development and extraction, would provide a stronger foundation for the formulation of supply shocks and their implications for behavior.

Heterogeneity and Market Identification

The only obvious recourse in addressing the heterogeneity problem is to make better use of the partially disaggregated data that are available, as, for example, in MacAvoy and Pindyck (1973, 1975), to better understand regional differences in behavior and the implications of shifting effort toward new regions (such as offshore and Rocky Mountain petroleum deposits). Efforts can be undertaken through case studies to look behind state and regional aggregates to investigate more useful aggregation measures; that is, measures that correspond to underlying spatial differences rather than arbitrary political boundaries.

On the conceptual side, it would be most useful to develop specifications of aggregate cost functions and decision rules that can be rationalized by hypotheses about individual behavior and about the distribution of individual characteristics in the industry. Building on the approach of Solow and Wan (1976) and Hartwick (1982), for example, it may be feasible to develop aggregate extraction cost specifications that combine a "decline curve" hypothesis of individual extraction with assumptions about the cross-sectional distribution of decline rate parameters and the distribution of resource deposits by size and grade. However, the inevitable question is whether enough data can be obtained to test these joint hypotheses.

Turning to other factors influencing observed market outcomes, the main task is to bridge the gap between a theory of individual decisions and data that reflect numerous interdependent influences. Because existing empirical studies have neglected identification issues, the seriousness of these issues remains a matter for conjecture. In view of the changing structure of markets for energy and other resources over the past decade, however, investigation of identification problems deserves a high priority in future research. Specific topics include a better understanding of resource demands, both conceptual and practical, with an emphasis on changing possibilities for substitution and structural changes in resource-using capital stocks; more extensive study of key input markets, such as rotary-rig drilling services in the case of petroleum; and development of measures for valuing discoveries and reserves where transactions involving these inputs are not observable.

As illustrated by the discussion of government intervention in the petroleum industry, changing regulatory policies can have a critical influence on resource markets, often by masking the underlying response of supply decisions to economic incentives. Distinguishing the effects of regulation from other influences requires explicit hypotheses about the regulatory process, supplier responses to controls, and suppliers' expectations of regulatory change. The rational expectations approach described in Sargent (1981) and Eckstein and Eichenbaum (1983b) offers the potential for resolving these and other market identification issues, but the limitations of the approach also must be kept in mind.

Externalities and Market Structure

The focus of our discussion of market failures is on conceptual issues encountered in applying models of individual strategic behavior to analyze market behavior. Despite their limited empirical content, dynamic game models can be useful in shedding light on how decisions in the presence of uncontrolled common property externalities or noncompetitive markets differ from responses under different market organizations. This information may be useful in suggesting the direction (if not the magnitude) of bias in empirical studies that ignore strategic responses. In addressing these questions, recent advances in game theory concerned with the specification of expectations in a dynamic context and with entry-deterring threats, such as the work of Friedman (1979a, 1979b), Dixit (1980), and Newbery (1981), can be applied to resource industries and possibly to OPEC decision making as well.

Common property externalities also should be investigated through case studies using data from the Texas Railroad Commission and other regulatory bodies. The case studies could combine fact-gathering with econometric analysis, using as a point of departure the empirically tractable models developed in Epple, Hansen, and Roberds (1983). Finally, an understanding of the structure and function of long-term contracts is a prerequisite for interpreting historical data and for addressing changes in contract terms and market organization.

FINAL COMMENT

The conceptual and practical problems posed in this chapter have a broader relevance than nonrenewable resource economics. By the same token, specialists in a variety of fields can contribute to understanding of resource supply behavior. The influx of rational expectations methods from macroeconomics to petroleum economics is an example of the potential for cross-fertilization from other specialties in advancing understanding of nonrenewable resource markets. The challenge remains to develop models that have empirical content.

References

Adelman, M. A. 1972. *The World Petroleum Market* (Baltimore, Johns Hopkins University Press for Resources for the Future).

______, and J. L. Paddock. 1978. "An Aggregate Model of Petroleum Production Capacity and Supply Forecasting," MIT World Oil Project Working Paper (Cambridge, Mass., MIT).

Adler, R. J., E. G. Cazalet, F. M. Haas, R. A. Marshalla, D. M. Nisbitt, and R. H. Phillips. 1979. *The DFI Generalized Equilibrium Modeling System* (Palo Alto, Decision Focus, Inc.).

Aivazian, V., and J. Callen. 1979. "A Note on the Economics of Exhaustible Resources," *Canadian Journal of Economics* vol. 12, no. 1 (February) pp. 83–89.

American Gas Association (AGA). 1973. *Total Energy Resources Analysis* (*TERA*). System Documentation Manual (Arlington, Va.).

______. 1977. *TERA*: *Total Energy Resource Analysis Model* vol. I, Offshore Gas and Oil Supply Model, AGA F30078 (Arlington, Va.).

______. 1979. *TERA*: *Total Resource Analysis Model*. Technical Documentation, vol. II, Onshore Gas and Oil Supply Model, AGA F30178 (Arlington, Va.).

American Petroleum Institute (API). 1976. *Standard Definitions for Petroleum Statistics*. Technical Report no. 1, 2nd ed. (Washington, D.C., American Petroleum Institute).

Arrow, K. J. 1962. "The Economic Implications of Learning by Doing," *Review of Economic Studies* vol. 29, no. 2 (June) pp. 155–172.

Arrow, K. J. 1971. *Essays in the Theory of Risk Bearing* (Chicago, Markham Publishing Company).

______, and S. Chang. 1982. "Optimal Pricing, Use, and Exploration of Uncertain Natural Resource Stocks," *Journal of Environmental Economics and Management* vol. 9, no. 1 (March) pp. 1–10.

Becker, R. A. 1980. "A Neoclassical Model of Optimal Capital Accumulation, Exhaustible Resource Extraction and Environmental Externalities," Economics Discussion Paper 80-16 (Bloomington, Indiana University).

Bellman, R. 1957. *Dynamic Programming* (Princeton, Princeton University Press).

Bertsekas, D. P. 1976. *Dynamic Programming and Stochastic Control* (New York, Academic Press).

Bohi, D. R. 1981. *Analyzing Demand Behavior: A Study of Energy Elasticities* (Baltimore, Johns Hopkins University Press for Resources for the Future).

______. 1983. "What Causes Oil Price Shocks?" Discussion Paper D-82S (Washington, D.C., Resources for the Future).

______, and M. Russell. 1978. *Limiting Oil Imports: An Economic History and Analysis* (Baltimore, Johns Hopkins University Press for Resources for the Future).

Bradley, P. 1967. *The Economics of Crude Petroleum Production* (Amsterdam, North-Holland).

Broadman, H. G. 1981. "Intraindustry Structure, Integration Strategies and Petroleum Firm Performance" (Ph.D. dissertation, Department of Economics, University of Michigan, Ann Arbor).

______, and W. D. Montgomery. 1983. *Natural Gas Markets After Deregulation* (Washington, D.C., Resources for the Future).

Bruno, M., and J. Sachs. 1982. "Input Price Shocks and the Slowdown in Economic Growth: The Case of UK Manufacturing," *Review of Economic Studies* vol. 49, no. 5 (Special Issue) pp. 677–706.

Burt, O., and R. Cummings. 1970. "Production and Investment in Natural Resource Industries," *American Economic Review* vol. 60, no. 4 (September) pp. 576–590.

Cagan, P. 1956. "The Monetary Dynamics of Hyperinflation," in M. Friedman, ed., *Studies in the Quantity Theory of Money* (Chicago, University of Chicago Press).

Clark, P., P. Coene, and D. Logan. 1981. "A Comparison of Ten U.S. Oil and Gas Supply Models," *Resources and Energy* vol. 3, no. 4 (December) pp. 297–335.

Cox, J. C., and A. W. Wright. 1976. "The Determinants of Investment in Petroleum Reserves and Their Implications for Public Policy," *American Economic Review* vol. 66, no. 1 (March) pp. 153–167.

Dasgupta, P., and G. Heal. 1974. "The Optimal Depletion of Exhaustible Resources," in *Review of Economic Studies Symposium on the Economics of Exhaustible Resources* (Edinburgh, Scotland, Longman Group Ltd.).

______, and ______. 1979. *Economic Theory and Exhaustible Resources* (Cambridge, James Nisbet and Co. and Cambridge University Press).

Davidson, P. 1963. "Public Policy Problems of the Domestic Crude Oil Industry," *American Economic Review* vol. 53, no. 1 (March) pp. 85–108.

Devarajan, S., and A. C. Fisher. 1981. "Hotelling's 'Economics of Exhaustible Resources': Fifty Years Later," *Journal of Economic Literature* vol. 19, no. 1 (March) pp. 65–73.

Dixit, A. 1980. "The Role of Investment in Entry-Deterrence," *Economic Journal* vol. 90, no. 1 (March) pp. 95–106.

Eckbo, P. L., H. D. Jacoby, and J. L. Smith. 1978. "Oil Supply Forecasting: A Disaggregated Process Approach," *Bell Journal of Economics* vol. 9, no. 1 (Spring) pp. 218–235.

Eckstein, Z., and M. S. Eichenbaum. 1983a. "Oil Supply Disruptions and the Optimal Tariff in a Dynamic Stochastic Equilibrium Model," Discussion Paper D-82F (Washington, D.C., Resources for the Future).

_____, and _____. 1983b. "Inventories and Quantity Constrained Equilibria in Regulated Markets," Discussion Paper D-82M (Washington, D.C., Resources for the Future).

Epple, D. N. 1975. *Petroleum Discoveries and Government Policy: An Econometric Study of Supply* (Cambridge, Mass., Ballinger Publishing Co.).

_____. 1983. "The Econometrics of Exhaustible Resource Supply: A Theory and an Application," Discussion Paper D-82R (Washington, D.C., Resources for the Future).

_____, and L. P. Hansen. 1979. "An Econometric Model of U.S. Petroleum Supply with Optimal Endogenous Depletion" (Pittsburgh, Carnegie-Mellon University, Graduate School of Industrial Administration).

_____, and _____. 1981. "An Econometric Framework for Modeling Exhaustible Resource Supply," in J. Ramsey, ed., *The Economics of Exploration for Energy* (Greenwich, JAI Press).

_____, _____, and W. Roberds. 1983. "Linear-Quadratic Games of Resource Depletion," Discussion Paper D-82Q (Washington, D.C., Resources for the Future).

Erickson, E., and R. M. Spann. 1971. "Supply Response in a Regulated Industry: The Case of Natural Gas," *Bell Journal of Economics* vol. 2, no. 1 (Spring) pp. 94–121.

Eswaran, M., and T. R. Lewis. 1981. "Production from Differentiated Exhaustible Resources in Oligopolistic Markets," Department of Economics, Resources Paper 63 (Vancouver, B.C., Canada, University of British Columbia).

_____, and _____. 1982. "Exhaustible Resources, Property Rights, and Alternative Equilibrium Concepts," Department of Economics, Discussion Paper 82-34 (Vancouver, B.C., Canada, University of British Columbia).

Eyssell, J. H. 1978. "The Supply Response of Crude Petroleum: New and Optimistic Results," *Business Economics* vol. 13, no. 3 (May) pp. 15–28.

Farrow, S. 1983. "An Empirical Method and Case Study to Test the Economic Efficiency of Extraction from a Stock Resource" (Pittsburgh, Carnegie-Mellon University, School of Urban and Public Affairs).

Fisher, A. C. 1982. *Resource and Environmental Economics* (Cambridge, England, Cambridge University Press).

Fisher, F. M. 1964. *Supply and Costs in the U.S. Petroleum Industry: Two Econometric Studies* (Baltimore, Johns Hopkins University Press for Resources for the Future).

Fisher, F. M. 1966. *The Identification Problem in Econometrics* (New York, McGraw-Hill).

______, and J. J. McGowan. 1983. "On the Misuse of Accounting Rates of Return to Infer Monopoly Profits," *American Economic Review* vol. 73, no. 1 (March) pp. 82–97.

Friedman, J. 1977. *Oligopoly and the Theory of Games* (Amsterdam, North-Holland).

______. 1979a. "Non-Cooperative Equilibria for Exit Supergames," *International Economic Review* vol. 20, no. 1 (February) pp. 147–156.

______. 1979b. "On Entry Preventing Behavior and Limit Price Models of Entry," *Applied Game Theory* (Wurzberg, Germany, Physica-Verlag).

______. 1983. "Low Information Nash Equilibria for Oligopolistic Markets," *Information Economics and Policy* vol. 1, no. 1, pp. 37–53.

Gilbert, R. J. 1978. "Dominant Firm Pricing Policy in a Market for an Exhaustible Resource," *Bell Journal of Economics* vol. 9, no. 2 (Autumn) pp. 385–395.

______, and S. M. Goldman. 1978. "Potential Competition and the Monopoly Price of an Exhaustible Resource," *Journal of Economic Theory* vol. 17, no. 1 (February) pp. 319–331.

Gilley, O. W., and G. V. Karels. 1981. "The Competitive Effect in Bonus Bidding: New Evidence," *Bell Journal of Economics* vol. 12, no. 2 (Autumn) pp. 637–648.

Granger, C. W. J., and P. Newbold. 1977. *Forecasting Economic Time Series* (New York, Academic Press).

Griffin, J. M., and D. Teece. 1982. *OPEC Behavior and World Oil Prices* (London, Allen & Unwin).

Grossman, S. J., R. E. Kihlstrom, and L. J. Mirman. 1977. "A Bayesian Approach to the Production of Information and Learning by Doing," *Review of Economic Studies* vol. 44, no. 3 (October) pp. 533–547.

Hansen, L. P., and T. J. Sargent. 1980. "Formulating and Estimating Dynamic Linear Rational Expectations Models," *Journal of Economic Dynamics and Control* vol. 2, no. 1, pp. 7–46.

______, and ______. 1981. "Linear Rational Expectations Models for Dynamically Interrelated Variables," in R. E. Lucas and T. J. Sargent, eds., *Rational Expectations and Econometric Practice* vol. 1 (Minneapolis, University of Minnesota Press).

Hartwick, J. M. 1978. "Exploitation of Many Deposits of an Exhaustible Resource," *Econometrica* vol. 46, no. 1 (January) pp. 201–217.

______. 1982. "Differential Resource Rents and the Two Theories of Non-Renewable Resource Valuation," *Resources and Energy* vol. 4, no. 3 (September) pp. 281–289.

Hausman, J. A. 1978. "Specification Tests in Econometrics," *Econometrica* vol. 46, no. 6 (November) pp. 1251–1272.

Heal, G. 1976. "The Relationship Between Price and Extraction Cost for a Resource with a Backstop Technology," *Bell Journal of Economics* vol. 7, no. 2 (Autumn) pp. 371–378.

______. 1982. "The Use of Common Property Resources," in V. K. Smith and

J. V. Krutilla, eds., *Explorations in Natural Resource Economics* (Baltimore, Johns Hopkins University Press for Resources for the Future).

Henry, C. 1974. "Investment Decisions Under Uncertainty: The 'Irreversibility Effect,'" *American Economic Review* vol. 64, no. 6 (December) pp. 1006–1012.

Herfindahl, O., and A. Kneese. 1974. *Economic Theory of Natural Resources* (Columbus, Charles E. Merrill).

Hnyilicza, E., and R. Pindyck. 1976. "Pricing Policies for a Two-Part Exhaustible Resource Cartel: The Case of OPEC," *European Economic Review* vol. 8, no. 3 (August) pp. 139–154.

Hotelling, H. 1931. "The Economics of Exhaustible Resources," *Journal of Political Economy* vol. 39, no. 2 (April) pp. 137–175.

Huntington, H. G. 1978. "Federal Price Regulation and the Supply of Natural Gas in a Segmented Field Market," *Land Economics* vol. 54, no. 3 (August) pp. 337–347.

Kamien, M. I., and N. L. Schwartz. 1969. "Induced Factor Augmenting Technical Progress from a Microeconomic Viewpoint," *Econometrica* vol. 37, no. 4 (October) pp. 668–684.

Karlin, S., and H. Taylor. 1975. *A First Course in Stochastic Processes* (New York, Academic Press).

Kaufman, G. 1963. *Statistical Decisions and Related Techniques in Oil and Gas Exploration* (Englewood Cliffs, N.J., Prentice-Hall).

Kemp, M., and N. Long. 1980a. "Toward a More General Theory of the Mining Firm," in their *Exhaustible Resources, Optimality, and Trade* (Amsterdam, North-Holland).

______, and ______. 1980b. "On Two Folk Theorems Concerning the Extraction of Exhaustible Resources," *Econometrica* vol. 48, no. 3 (April) pp. 663–673.

Khazzoom, J. D. 1971. "The FPC Staff's Econometric Model of Natural Gas Supply in the United States," *Bell Journal of Economics* vol. 2, no. 1 (Spring) pp. 51–93.

Kim, Y. Y., and R. G. Thompson. 1978. *New Oil and Gas Supplies in the Lower 48 States* (Houston, Gulf Publishing Co.).

Kuller, R., and R. Cummings. 1974. "An Economic Model of Production and Investment for Petroleum Reservoirs," *American Economic Review* vol. 64, no. 1 (March) pp. 66–79.

Lee, D. R. 1978. "Price Controls, Binding Constraints, and Intertemporal Economic Decision Making," *Journal of Political Economy* vol. 86, no. 2 (April) pp. 293–302.

______. 1979. "Price Controls on Non-Renewable Resources: An Intertemporal Analysis," *Southern Economic Journal* vol. 46, no. 1 (July) pp. 179–188.

______. 1981. "Monopoly, Price Controls, and the Exploitation of Nonrenewable Resources," *Journal of Energy and Development* vol. 7, no. 1 (Autumn) pp. 111–120.

Levhari, D., and N. Liviatan. 1977. "Notes on Hotelling's Economics of Exhaustible Resources," *Canadian Journal of Economics* vol. 10, no. 2 (May) pp. 177–192.

Liu, P. T., and J. G. Sutinen. 1982. "On the Behavior of Optimal Exploration and Extraction Rates for Non-Renewable Resource Stocks," *Resources and Energy* vol. 4, no. 2 (June) pp. 145–162.

Lovejoy, W. F., and P. T. Homan. 1965. *Methods of Estimating Reserves of Crude Oil, Natural Gas, and Natural Gas Liquids* (Baltimore, Johns Hopkins University Press for Resources for the Future).

Lucas, R. E., Jr. 1976. "Econometric Policy Evaluation: A Critique," in K. Brunner and A. Meltzer, eds., *The Phillips Curve and Labor Markets*, supplementary series to the *Journal of Monetary Economics* vol. 1 (Amsterdam, North-Holland).

MacAvoy, P., and R. S. Pindyck. 1973. "Alternative Regulatory Policies for Dealing with the Natural Gas Shortage," *Bell Journal of Economics* vol. 4, no. 2 (Autumn) pp. 454–498.

______, and ______. 1975. *The Economics of the Natural Gas Shortage 1960–1980* (Amsterdam, North-Holland).

Maddock, R., and M. Carter. 1982. "A Child's Guide to Rational Expectations," *Journal of Economic Literature* vol. 20, no. 1 (March) pp. 39–51.

Mancke, R. B. 1970. "The Long-Run Supply Curve of Crude Oil Produced in the United States," *The Antitrust Bulletin* vol. 15 (Winter) pp. 727–756.

McDonald, S. L. 1971. *Petroleum Conservation in the United States: An Economic Analysis* (Baltimore, Johns Hopkins University Press for Resources for the Future).

______. 1979. *The Leasing of Federal Lands for Fossil Fuels Production* (Baltimore, Johns Hopkins University Press for Resources for the Future).

Miller, C. L. 1982. "A Reaction Function Equilibrium for a Simple Dynamic Duopoly Model with Random Demand," Department of Political Economy Working Paper 94 (Baltimore, Johns Hopkins University).

Modiano, E., and J. Shapiro. 1980. "A Dynamic Optimization Model of Depletable Resources," *Bell Journal of Economics* vol. 11, no. 1 (Spring) pp. 212–236.

Montgomery, W. D. 1977. "The Transition to Uncontrolled Crude Oil Prices," Social Science Working Paper 186 (Pasadena, California Institute of Technology).

Moran, T. H. 1981. "Modeling OPEC Behavior: Economic and Political Alternatives," *International Organization* vol. 35, no. 2 (Spring) pp. 241–272.

Nehring, R. 1981. *The Discovery of Significant Oil and Gas Fields in the United States*. Rand Report published for the U.S. Departments of Interior and Energy (R-2654/1 USGS/DOE).

Newbery, D. M. G. 1981. "Oil Prices, Cartels, and the Problem of Dynamic Inconsistency," *Economy Journal* vol. 91, no. 3 (September) pp. 617–646.

Peterson, F. M. 1978. "A Model of Mining and Exploring for Exhaustible Resources," *Journal of Environmental Economics and Management* vol. 5, no. 3 (September) pp. 236–251.

______, and A. C. Fisher. 1977. "The Exploitation of Extractive Resources: A Survey," *Economic Journal* vol. 87, no. 4 (December) pp. 681–721.

Pindyck, R. S. 1977. "Cartel Pricing and the Structure of the World Bauxite Market," *Bell Journal of Economics* vol. 8, no. 2 (Autumn) pp. 343–360.

______. 1978a. "Gains to Producers from the Cartelization of Exhaustible

Resources," *Review of Economics and Statistics* vol. 60, no. 2 (May) pp. 238–251.

______. 1978b. "Higher Energy Prices and the Supply of Natural Gas," *Energy Systems and Policy* vol. 2, no. 2 (Spring) pp. 177–199.

______. 1978c. "The Optimal Exploration and Production of Nonrenewable Resources," *Journal of Political Economy* vol. 86, no. 5 (October) pp. 841–861.

______. 1980. "Uncertainty and Exhaustible Resource Markets," *Journal of Political Economy* vol. 88, no. 6 (December) pp. 1203–1225.

______. 1981a. "The Optimal Production of an Exhaustible Resource When Price Is Exogenous and Stochastic," *Scandinavian Journal of Economics* vol. 83, no. 2 (Spring) pp. 277–289.

______. 1981b. "The Optimal Phasing of Phased Deregulation," MIT Energy Laboratory Discussion Paper 3 (Cambridge, Mass., MIT).

______. 1981c. "Models of Resource Markets and the Explanation of Resource Price Behavior," *Energy Economics* vol. 3, no. 3 (July) pp. 130–139.

______. 1982. "Jointly Produced Exhaustible Resources," *Journal of Environmental Economics and Management* vol. 9, no. 4 (December) pp. 291–303.

______, and D. L. Rubinfeld. 1976. *Econometric Models and Economic Forecasts* (New York, McGraw-Hill).

Plosser, C. I., G. W. Schwerf, and H. White. 1982. "Differencing as a Test of Specification," *International Economic Review* vol. 23, no. 3 (October) pp. 535–552.

Ramsey, J. B. 1980. "The Economics of Oil Exploration: A Probability-of-Ruin Approach," *Energy Economics* vol. 2, no. 1 (January) pp. 14–30.

Reinganum, J., and N. Stokey. 1981. "Oligopoly Extraction of a Nonrenewable, Common Property Resource: The Importance of the Period of Commitment in Dynamic Games," Social Science Working Paper 377 (Pasadena, California Institute of Technology).

Rice, P., and V. K. Smith. 1977. "An Econometric Model of the Petroleum Industry," *Journal of Econometrics* vol. 6, no. 3 (November) pp. 263–287.

Roberds, W. 1982. "A Dominant Firm Model of an Exhaustible Resource Industry: Theory and Estimation" (Providence, Brown University, Department of Economics).

Russell, M. 1972. "Producer Regulation for the 1970s," in K. Brown, ed., *Regulation of the Natural Gas Producing Industry* (Washington, D.C., Resources for the Future).

Salant, S. W. 1976. "Exhaustible Resources and Industrial Structure: A Nash-Cournot Approach to the World Oil Market," *Journal of Political Economy* vol. 84, no. 5 (October) pp. 1079–1093.

______. 1982. "Imperfect Competition in the International Energy Market: A Computerized Nash-Cournot Model," *Operations Research* vol. 30, no. 2 (March–April) pp. 252–280.

Sargent, T. J. 1979. *Macroeconomic Theory* (New York, Academic Press).

______. 1981. "Interpreting Economic Time Series," *Journal of Political Economy* vol. 89, no. 2 (April) pp. 213–248.

Schmidt, P. 1976. *Econometrics* (New York, Marcel Dekker).

Slade, M. E. 1982a. "Trends in Natural-Resource Commodity Prices: An Analysis of the Time Domain," *Journal of Environmental Economics and Management* vol. 9, no. 2 (June) pp. 122–137.

______. 1982b. "Cycles in Natural-Resource Commodity Prices: An Analysis of the Frequency Domain," *Journal of Environmental Economics and Management* vol. 9, no. 2 (June) pp. 138–148.

Smith, V. K., ed. 1979. *Scarcity and Growth Reconsidered* (Baltimore, Johns Hopkins University Press for Resources for the Future).

______. 1981. "The Empirical Relevance of Hotelling's Model for Natural Resources," *Resources and Energy* vol. 3, no. 2 (October) pp. 105–117.

______, and J. V. Krutilla. 1982. *Explorations in Natural Resource Economics* (Baltimore, Johns Hopkins University Press for Resources for the Future).

Solow, R. M. 1974. "The Economics of Resources or the Resources of Economics," *American Economic Review* vol. 64, no. 2 (May) pp. 1–14.

______, and F. Y. Wan. 1976. "Extraction Costs in the Theory of Exhaustible Resources," *Bell Journal of Economics* vol. 7, no. 2 (Autumn) pp. 359–370.

Spann, R. M., and E. W. Erickson. 1973. "Joint Costs and Separability in Oil and Gas Separation," in M. F. Searl, ed., *Energy Modeling: Art, Science, Practice* (Washington, D.C., Resources for the Future).

Stauffer, T. R. 1971. "The Measurement of Corporate Rates of Return: A Generalized Formulation," *Bell Journal of Economics* vol. 2, no. 2 (Autumn) pp. 434–469.

Stiglitz, J. 1974a. "Growth with Exhaustible Natural Resources: Efficient and Optimal Growth Paths," in *Review of Economic Studies Symposium on the Economics of Exhaustible Resources* (Edinburgh, Scotland, Longman Group Ltd.).

______. 1974b. "Growth with Exhaustible Natural Resources: The Competitive Economy," in *Review of Economic Studies Symposium on the Economics of Exhaustible Resources* (Edinburgh, Scotland, Longman Group Ltd.).

______. 1975. "The Efficiency of Market Prices in Long-Run Allocations in the Oil Industry," in G. Brannon, ed., *Studies in Energy Tax Policy* (Cambridge, Mass., Ballinger).

Stollery, K. R. 1983. "Mineral Depletion with Cost as the Extraction Limit: A Model Applied to the Behavior of Prices in the Nickel Industry," *Journal of Environmental Economics and Management* vol. 10, no. 2 (June) pp. 151–165.

Theil, H. 1971. *Principles of Econometrics* (New York, Wiley).

Toman, M. A. 1982a. "A Noncooperative Equilibrium for State Dependent Supergames," Discussion Paper D-82I (Washington, D.C., Resources for the Future).

______. 1982b. "The Theory of Nonrenewable Resource Exploitation: Existence of an Optimal Plan," Discussion Paper D-98 (Washington, D.C., Resources for the Future).

______. 1982c. "Perfect Foresight Market Equilibrium with a Nonrenewable Resource," Discussion Paper D-99 (Washington, D.C., Resources for the Future).

______. 1983a. "A Diagrammatic Exposition of Nonrenewable Resource Production and Exploration," Discussion Paper D-111 (Washington, D.C., Resources for the Future).

______. 1983b. "Common Property Resource Externalities and Free Entry," Discussion Paper D-112 (Washington, D.C., Resources for the Future).

______. 1983c. "Common Property Resource Externalities and Entry Deterrence," in R. L. Itteilag, ed., *Government and Energy Policy* (Alexandria, Va., International Association of Energy Economists).

Tourinho, O. A. 1979. "The Option Value of Reserves of Natural Resources," Finance Working Paper 94 (Berkeley, University of California, Graduate School of Business Administration).

Treadway, A. B. 1970. "Adjustment Costs and Variable Inputs in the Theory of the Competitive Firm," *Journal of Economic Theory* vol. 2, no. 4 (December) pp. 329–347.

Uhler, R. S. 1979. "The Rate of Petroleum Exploration and Extraction," in R. S. Pindyck, ed., *Advances in the Economics of Energy and Resources* vol. 2 (Greenwich, JAI Press).

U.S. Department of Energy (DOE). 1978. *Midterm Oil and Gas Supply Modeling System: Methodology Description*. Technical Memorandum TM/ES/79-05 (Washington, D.C., Department of Energy).

______. 1979a. *Alaskan Hydrocarbon Supply Model: Methodology Description*. Technical Memorandum TM/ES/79-09 (Washington, D.C., Department of Energy).

______. 1979b. *Enhanced Oil Recovery Model: Methodology Description*. Technical Memorandum TM/ES/79-7 (Washington, D.C., Department of Energy).

Verleger, P. K. 1982. "The Determinants of Official OPEC Crude Prices," *Review of Economics and Statistics* vol. 64, no. 2 (May) pp. 171–183.

Wallis, K. F. 1980. "Econometric Implications of the Rational Expectations Hypothesis," *Econometrica* vol. 48, no. 1 (January) pp. 49–74.

Zabel, E. 1971. "Risk and the Competitive Firm," *Journal of Economic Theory* vol. 3, no. 1 (March) pp. 109–133.

______. 1982. "Trading in a Monopolistic Market," Discussion Paper 50 (Gainesville, University of Florida, Center for Econometrics and Decision Sciences).

Zimmerman, M. B. 1975. "Long-Run Mineral Supply: The Case of Coal in the United States" (Ph.D. dissertation, Massachusetts Institute of Technology, Cambridge, Mass.).

______. 1977. "Modeling Depletion in the Mineral Industry: The Case of Coal," *Bell Journal of Economics* vol. 8, no. 1 (Spring) pp. 41–65.

Index

ABOUT THE AUTHORS

Douglas R. Bohi is a Senior Fellow at the Center for Energy Policy Research of Resources for the Future, Inc. He was formerly Chairman of the Department of Economics at Southern Illinois University, Carbondale. He received his Ph.D. in economics from Washington State University.

Michael A. Toman is a Fellow at the Center for Energy Policy Research of Resources for the Future, Inc. He received the Ph.D. in economics from the University of Rochester.